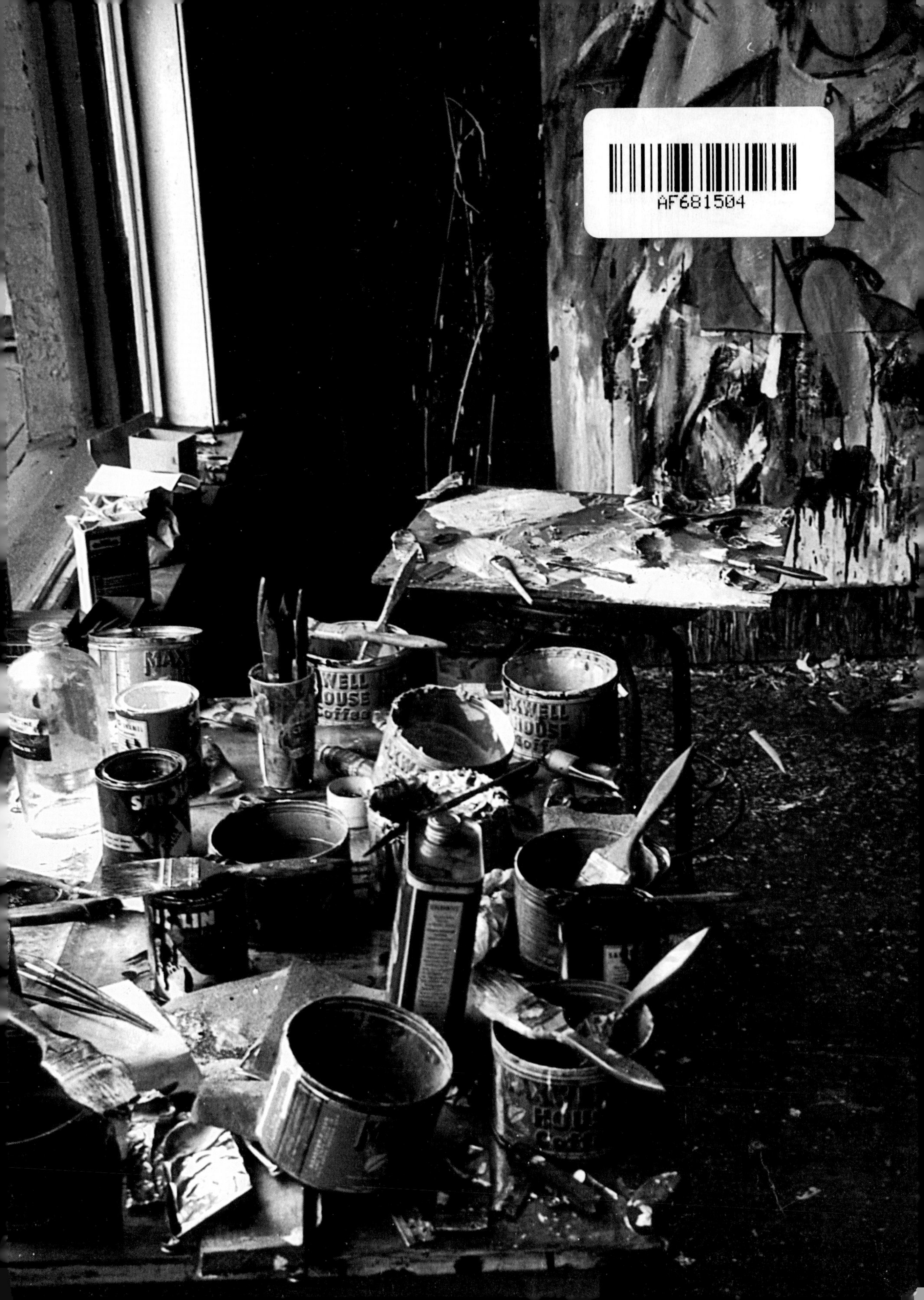

AF681504

SUSAN F. LAKE

WILLEM DE KOONING
THE ARTIST'S MATERIALS

With technical analysis by
Susan F. Lake and Michael Schilling

The Getty Conservation Institute
Los Angeles

The Getty Conservation Institute

Timothy P. Whalen, *Director*
Jeanne Marie Teutonico, *Associate Director, Programs*

The Getty Conservation Institute works internationally to advance conservation practice in the visual arts—broadly interpreted to include objects, collections, architecture, and sites. The Institute serves the conservation community through scientific research, education and training, model field projects, and the dissemination of the results of both its own work and the work of others in the field. In all its endeavors, the GCI focuses on the creation and delivery of knowledge that will benefit the professionals and organizations responsible for the conservation of the world's cultural heritage.

© 2010 J. Paul Getty Trust

Published by the Getty Conservation Institute
Getty Publications
1200 Getty Center Drive, Suite 500
Los Angeles, California 90049-1682
www.gettypublications.org

Gregory M. Britton, *Publisher*

Tobi Levenberg Kaplan
and Beatrice Hohenegger, *Editors*
Robin Ray, *Manuscript Editor*
Suzanne Watson, *Production Coordinator*
Jeffrey Cohen, *Designer*

Printed in China through Asia Pacific Offset, Inc., on Gold East matte

FRONT COVER: Detail of *Woman, Sag Harbor,* 1964. Oil and charcoal on wood, 203.1 × 91.2 cm (80 × 36 in.). Hirshhorn Museum and Sculpture Garden, Smithsonian Institution (66.1209). Gift of Joseph H. Hirshhorn, 1966. © 2010 The Willem de Kooning Foundation/Artists Rights Society, New York.

PAGE I: Interior of de Kooning's studio, New York, 1950. Photo © 2010 Estate of Rudy Burckhardt / Artists Rights Society (ARS), New York. Artwork © 2010 The Willem de Kooning Foundation/ Artists Rights Society, New York.

PAGE VIII: De Kooning in his studio, Springs, East Hampton, 1971. Artwork © 2010 The Willem de Kooning Foundation/Artists Rights Society, New York. Photo: Dan Budnik.

PAGE 4: De Kooning examining his drawings, Springs, East Hampton, 1971. Artwork © 2010 The Willem de Kooning Foundation/Artists Rights Society, New York. Photo: Dan Budnik.

BACK COVER: De Kooning in his Broadway studio, New York, 1962. Photo © 2010 Estate of Rudy Burckhardt / Artists Rights Society (ARS), New York. Artwork © 2010 The Willem de Kooning Foundation/Artists Rights Society, New York.

Every effort has been made to contact the copyright holders of the materials in this book to obtain permission to publish. Any omissions will be corrected in future editions if the publisher is contacted in writing.

Library of Congress
Cataloging-in-Publication Data

Lake, Susan.
Willem de Kooning : the artist's materials / Susan F. Lake ; with technical analyses by Susan Lake and Michael Schilling.
p. cm.
Includes bibliographical references and index
ISBN 978-1-60606-021-6 (pbk.)
1. De Kooning, Willem, 1904–1997.
2. Painting, Modern—20th century—Technique.
3. Paint materials—Analysis. 4. Artists' materials —Analysis. I. De Kooning, Willem, 1904–1997.
II. Schilling, Michael (Michael Roy) III. Title.
IV. Title: Artist's materials.
ND237.D334L35 2009
759.13—dc22

2009045607

Contents

Contents, cont'd

Foreword

Timothy P. Whalen

Director,
Getty Conservation Institute

Willem de Kooning, one of the most influential painters of the twentieth century, was like many artists of his generation who adopted a number of unorthodox approaches to the mixing and application of their paints. The unique materials, methods, and techniques de Kooning used to create his paintings all have important implications for their preservation.

The Getty Conservation Institute (GCI) is working to address the challenges of conserving art of the twentieth and twenty-first centuries through its Modern and Contemporary Art Research Initiative. Although much of our activity in this area has involved the scientific study of new materials being used by artists (specifically, analysis of modern paints) and research into the effects of conservation treatments on those materials, we also recognize the very real need for effective avenues for dissemination and information sharing. For example, the GCI, the Tate in London, and the National Gallery of Art in Washington, D.C., hosted the "Modern Paints Uncovered Symposium" in 2006 (proceedings published in 2007), and the GCI and the Getty Research Institute (GRI) sponsored the "Object in Transition" conference in 2008.

Because conservators must understand the materials used on works of art in order to undertake safe and successful treatments, a series of publications focusing on the study and conservation of modern artists' materials has been conceived. *Willem de Kooning: The Artist's Materials*, by Susan F. Lake, begins the series.

Susan Lake, head of collection management and chief conservator at the Hirshhorn Museum and Sculpture Garden in Washington, D.C., has written this book based on her many years of work researching the painting materials of de Kooning, much of it in collaboration with Michael Schilling, senior scientist at the GCI. Susan is well known and respected in the field for her research interest in American Abstract Expressionist painters and the conservation of modern art materials and we have enjoyed several collaborations with her over the years.

De Kooning's working methods have long caused speculation, rumor, and debate among conservators and art historians, primarily on the basis of visual inspection of the paintings and anecdotal accounts rather than rigorous technical analysis. Lake's work constitutes the first systematic study of de Kooning's creative process that uses comprehensive scientific examinations of the artist's pigments, binders, and supports. The research informs both art historical interpretations of and conservation strategies for de Kooning's paintings.

I am thrilled that this work can now be shared with the field and am confident that it will be essential reading for anyone involved with the study or conservation of Willem de Kooning's paintings and a wide range of modern and contemporary art.

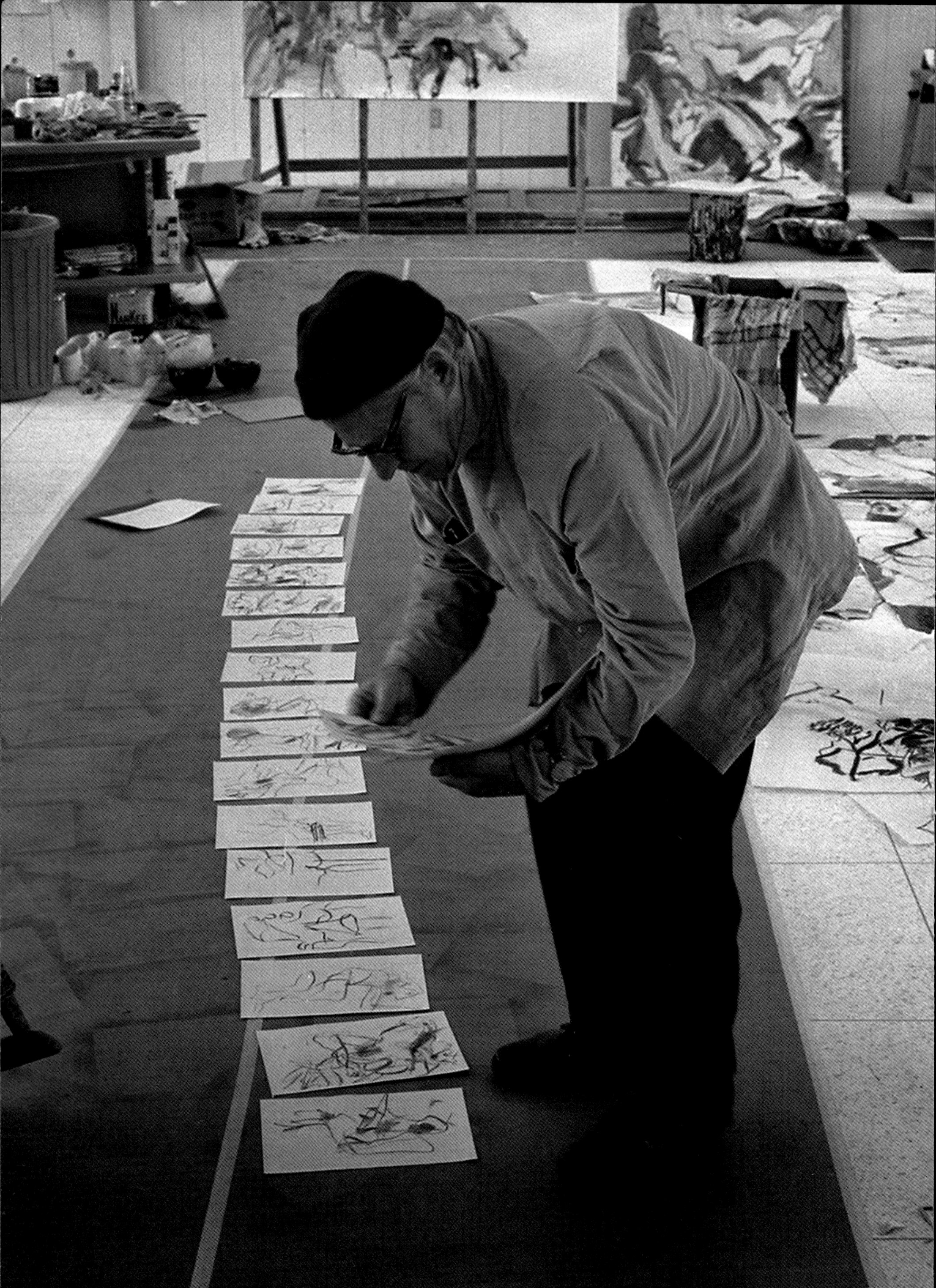

Introduction

Willem de Kooning (1904–1997), one of the major painters of the twentieth century, is widely known for his masterful yet highly idiosyncratic working methods. Less recognized is the extent to which the artist employed unconventional as well as traditional materials in his paintings to achieve visual effects that reflected his surroundings. Unlike many painters of the Abstract Expressionist movement, de Kooning developed several summary or signature styles; no single image represents his style or characterizes his career as do, for example, the poured paintings of Jackson Pollock and the abstracted topographies of Clyfford Still. Rather, de Kooning was a painter who mastered the use of art and non-art materials, changing his style and materials to evoke his various immediate environment.

This book is an expansion of the author's Ph.D. dissertation, "The Relationship between Style and Technical Procedure: Willem de Kooning's Paintings of the Late 1940s and 1960s" (University of Delaware, 1999). The study was catalyzed, in part, by two major exhibitions in Washington, D.C., during the 1990s: *Willem de Kooning from the Hirshhorn Museum Collection* (1993) and *Willem de Kooning: Paintings* (1994), organized by the National Gallery. Both installations allowed the author to closely examine de Kooning's paintings and to sample many of the seminal works of the artist's oeuvre. While the paintings included in this study are owned by various public and private collections in the United States and Europe, the majority of those examined and analyzed in depth are in the Hirshhorn Museum and Sculpture Garden collection. No other public repository of de Kooning's painting approaches the size and chronological scope of the Hirshhorn collection. Access to these works, which range in date from 1939 to 1985, provided a unique opportunity to better understand de Kooning's working methods and his motivations as they evolved over time.

This technical study focuses on a seminal period in the complicated evolution of Willem de Kooning's painting style: the late 1940s through the 1960s, integrating art historical resources with the results of technical examination and scientific analysis. Medium analysis from four works from the 1970s, carried out by Michael Schilling, is also included (Chapter 4), as many of these paintings are similar in appearance and condition to those of the previous decade. The author undertook pigment and medium analysis of thirteen paintings executed by de Kooning while he lived in Manhattan and twelve after he moved to Long Island. The results reveal the extent to which the artist persisted in inventing new ways to manipulate his paints and to exploit the properties of various supports, grounds, and paints.

De Kooning belonged to the community of artists that radically transformed American art during World War II and in its immediate aftermath. Abstract Expressionism, or the New York School, as the "movement" has alternatively been called, consisted of a loosely affiliated group of artists from disparate backgrounds, each of whom, in the 1940s, began to develop an individual style as a means of self-expression. Unlike artists in other movements, the Abstract Expressionists did not share a collective style or common thematic concerns that were apparent to the viewing public. They did, however, share a belief that a work of art did not need to narrate or illustrate to be an effective means of expression.

Willem de Kooning was, to an extent, rare among the Abstract Expressionist painters, largely inspired by the circumstances of his everyday life. Over the years, scholars have identified many possible sources for de Kooning's imagery and style, from old-master paintings to Picasso to commercial images found in advertisements and fashion magazines (Sylvester 1994, 15–31). While these connections make sense visually and historically, non-artistic influences on de Kooning's aesthetics were equally profound. The artist often affirmed that his immediate environment was a major resource and that his surroundings permeated his work (Hunter 1975, 70; "Inner Monologue" 1959, 3–8; Zilczer 1993a). Two major changes in location greatly affected the course of his art. First, in 1926, he emigrated from Rotterdam to New York City, and during the 1940s and 1950s Manhattan became a source of inspiration to his art. Then, in 1963, he moved permanently to Springs, Long Island, a small seaside community just outside the town of East Hampton. That environment, surrounded by water and bathed in the North Atlantic light, formed the basis of his paintings thereafter.

In a broad sense, the current research project attempts to determine the degree to which the artist's important shifts in style corresponded to changes in his technical procedures. *Woman*, painted in 1948 (see fig. 20), embodies the type of painting that de Kooning produced from the late 1940s into the 1950s. *Woman, Sag Harbor* (1964; see fig. 41) typifies the artist's treatment of the figure after 1963. Both works depict a female figure in a landscape setting composed of broad paint strokes, which integrate the figure into the background. Likewise, both subjects are portrayed as rather garish nude or nearly nude females, with toothy grins and bulging eyes and breasts. Despite their similarities, the two paintings have obvious differences. In the earlier work, de Kooning limited his range of colors, with black and white predominating, punctuated by strident patches of pink, blue, and yellow. In *Woman, Sag Harbor*, his palette had shifted to a narrow range of pinks, oranges, and yellows. Likewise, the more angular structure and the sharp, incisive contours of the earlier image contrast strongly with the softer, sweeping curves of the later painting. Lines drawn in black paint and charcoal, an important feature in *Woman* (1948), are not a prominent feature in *Woman, Sag Harbor*. Rather, the later painting evidences the increasing painterliness of his oeuvre. Probably the most distinctive aspect of this change in style is the difference in the surface textures of the two paintings. While the paint of *Woman* is matte, chunky, and coarse, that of *Woman, Sag Harbor* is unctuous and puckered, and it appears to glisten with moisture.

Changes in artistic style can be observed and described, but they are difficult to quantify. Art historical research can single out an element new to the artist's work and suggest possible sources of inspiration, such as other works of art or contact with fellow artists; yet, without documentary evidence of the artist's conscious choice in materials—and many modern artists are hesitant to discuss their painting materials—it is difficult to establish whether a change was deliberate or reflected an unconscious appropriation. Technical examination of the materials and techniques of paintings offers another avenue for research. It yields an additional body of evidence that, on occasion, can pinpoint artistic choices. If technical study establishes that an artist varied his/her practice in a specific way—by choosing one type of paint over another, or by altering the way the paint is applied—this is evidence of conscious artistic choice and deliberate innovation.

With the goal of examining the degree to which changes in style correspond to changes in painting practice, this technical study focuses on a critical period in de Kooning's career—the late 1940s through the 1960s.[1] Chapter 1 examines the arc of this career and his practices as discussed in previous literature. Chapters 2 and 3 discuss in some detail seven paintings from the late 1940s through the mid-1950s, and six from the 1960s. Chapter 4 presents medium analysis and a discussion of condition issues for sixteen works executed in the 1960s and 1970s, with methods and quantitative findings given in Appendixes 1 and 2. The study reveals that while de Kooning maintained certain lifelong painting practices, he selectively modified his materials and methods from decade to decade to produce paintings that appear significantly different from one another. The artist's selection of supports, grounds, and paint structure, including both pigments and binding media, has important implications for the surface textures of the pictures and their overall tonality.

In a now-famous statement of 1947, Jackson Pollock observed that a painting had "a life of its own" that would come through by means of the artist's handling of the materials. Likewise, de Kooning asserted that he never worked with the idea of making a perfect painting, but rather to see how far he could go (O'Connor and Thaw 1978, 241, D71; Sylvester 1963, 47). What these statements have in common is the artists' conviction that the materials and methods of painting served as expressive agents that had the power to convey messages, rather than acting merely as functional tools in the creative process. These artists believed that the times not only mandated a new way of painting, but that new materials had to be employed for these new methods. The transformation of American painting of this period is inextricably bound to the methods and the materials the artists used to make them.

Unlike most of the American Abstract Expressionist painters, de Kooning brought to his craft a rigorous European training. In his native Netherlands he apprenticed to a commercial design and decorating firm, and he attended the Rotterdam Academy of Fine Arts and Techniques. Once he arrived in the United States, he supported himself in lean times as a house painter and with a variety of commercial art jobs. As this book will show, both his training and his work experience gave him a respect for materials and skilled craftsmanship that was to be a fundamental influence upon his art throughout his career.

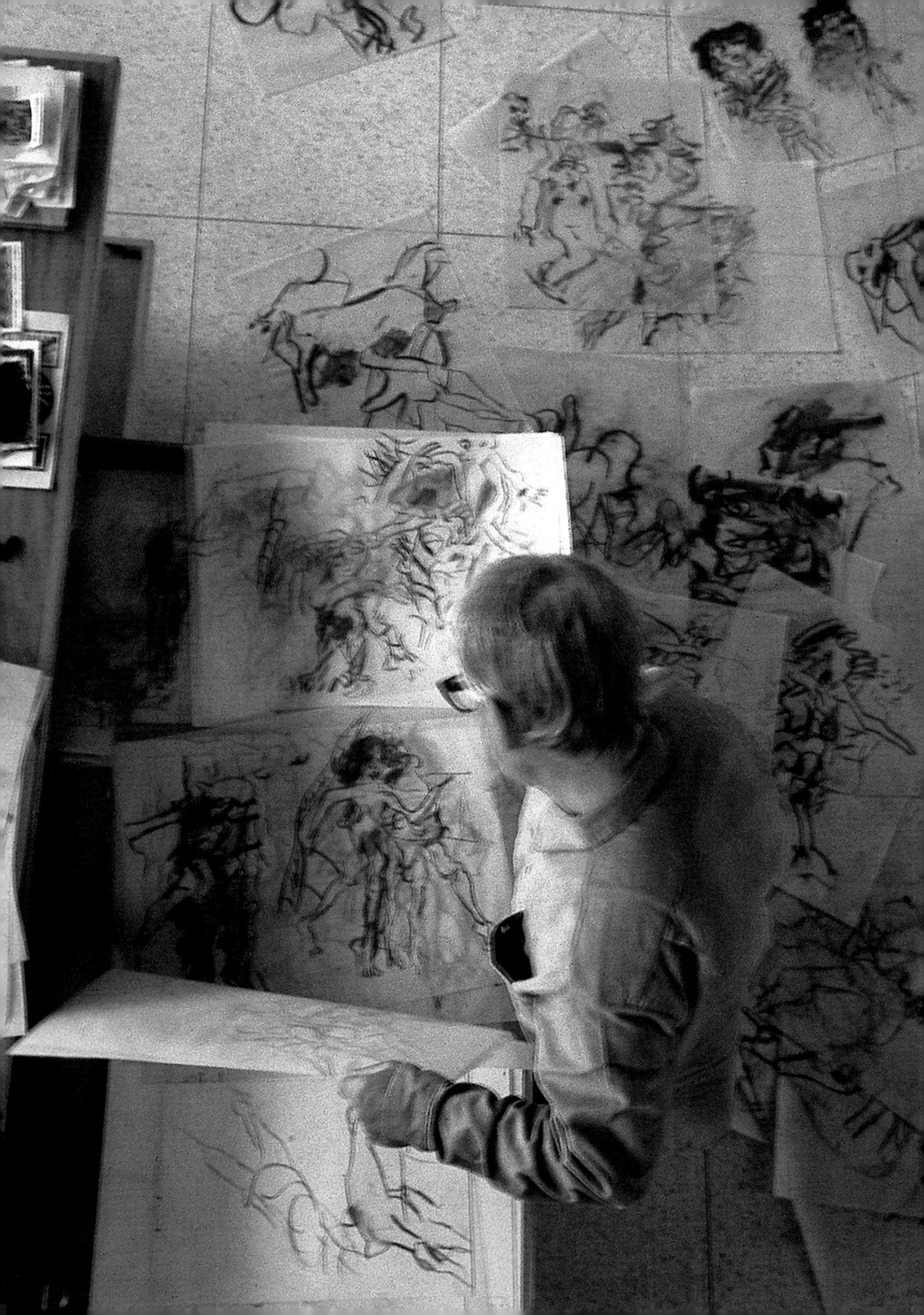

1 De Kooning in Perspective

Biographical Overview

From Holland to America

Willem de Kooning was born on April 24, 1904, in a rundown working-class district of Rotterdam North. As a child, he was exposed to poverty and had a difficult family life. His father, Leendert de Kooning, owned a wholesale beer and beverage distribution company; his mother, Cornelia Nobel, worked as a waitress in a bar that catered to seamen. They divorced when Willem was three years old. Willem and his older sister, Maria Cornelia, stayed with their mother, although Willem lived for a time with his father when his mother remarried (Wolfe 1996, 30–35; Stevens and Swan 2004, 9–10).

Like most children of working-class families, de Kooning went to free neighborhood schools for six years and then, at the age of twelve, left school to begin an apprenticeship. At school, de Kooning's teachers had noticed his talent for drawing and encouraged his mother and stepfather to direct his training toward a job as a craftsman. He began as an apprentice with a prestigious commercial design and decorating firm run by two brothers, Jan and Jaap Gidding. For the first several years of his apprenticeship, de Kooning ran errands, cleaned the workshop spaces, and learned the basics of art: grinding pigments and mixing paint. Later he learned to marbleize, wood-grain, and gild surfaces. "We did a lot of hotel lobbies and that sort of thing," he later recalled (Wilkie 2005, 25).

A year after he joined the decorating firm, de Kooning enrolled at the Rotterdam Academy of Fine Arts and Techniques, where he attended evening classes intermittently over the following eight years. Its rigorous curriculum combined academic classes with practical training in the modern industrial sciences and instruction in the guild and craft traditions. De Kooning attended lectures on color principles, art theory, and art history. He was trained in mechanical construction and perspective and learned the fundamentals of composition. He drew and painted from still-life arrangements and from classical plaster casts. Additionally, he learned the applied arts and guild techniques of stenciling, lettering, and sign painting (T. Hess 1968, 13–14; Stevens and Swan 2004, 3–45). His experience at the Academy instilled in him an abiding admiration for the humanistic traditions of Western art, and he was always justly proud of the technical training that he had received in the commercial and decorative arts and in the guild traditions. Both his academic education and his technical training gave him a fundamental knowledge of materials, a respect for skilled craftsmanship, and an enthusiasm for the popular and applied arts, which were to have a profound influence upon his art throughout his career.

Once de Kooning's apprenticeship was completed, he took various freelance jobs in the commercial art world. At eighteen, he still did not aspire to become a serious artist. He was a working-class boy who had to make a living. Moreover, as he later recounted, "When we went to the Academy—doing painting, decorating, making a living—young artists were not interested in painting *per se*. We used to call that 'good for men with beards.' And the idea of a palette, with colors on it, was rather silly. At the time we were influenced by the De Stijl group" (Sylvester 1963, 45). Eventually he was hired by a large Rotterdam department store to paint signs and arrange window displays. Working under Bernard Romein, the art director of the store, de Kooning developed an interest in avant-garde art movements. According to de Kooning, Romein was much more than a commercial artist. He was a true artist, "like a rebel and through him I got to know the work of Frank Lloyd Wright, Piet Mondrian and Walt Whitman" (Bourdrez 1950, 8).

As de Kooning approached twenty, he became increasingly frustrated with the course that his life was taking. He was weary of the Academy's rigid structure and became disenchanted with the provincial Dutch bias against modern design. With friends, he made several extended trips to Belgium. Initially he supported himself with a variety of commercial odd jobs, anything to make ends meet. Later, he was hired as a "painter-decorator" by a Brussels decorating business to execute impressionist-style landscapes that were sold through local department stores (Stevens and Swan 2004, 50). When he returned to Rotterdam in early 1925, he lived as a bohemian, on the margins of society and in genuine poverty. By this time, he was almost completely estranged from his family and had no steady source of income. It was at this point that he began to scheme of ways to go to America.

De Kooning later told Abram Lerner, then director of the Hirshhorn Museum, that since childhood America had epitomized the modern. He felt that Holland was small and too limited in opportunities. He loved American jazz and the photographs of girls in magazine advertisements; he was:

> wild for American illustrators..., having heard that successful illustrators in America were so affluent they worked only a few hours and played tennis every day. It was his ambition to become an illustrator and make his fortune. He read romantic novels of the old west, saw all the American movies, and was happy to believe everything (Lerner 1977, 334).[2]

After several failed attempts, de Kooning finally left the Netherlands with his friend Leo Cohan in 1926 as a stowaway on the *S.S. Shelley*, headed for Newport News, Virginia. Although they knew little English, they eventually made their way to New York through a network of Dutch friends. De Kooning rented a room temporarily at the Dutch Seaman's Home in Hoboken, N.J., and found work almost immediately as a house painter. On the job, he was impressed with the American workers' efficiency and their good-quality brushes. He later also recalled that he learned a lot about how to mix paint with water and oil (Rodgers 1978, 20).

In 1927, de Kooning moved from Hoboken to a studio apartment on West Forty-second Street in Manhattan, the first of a series of New York apartments and lofts, for the most part in Greenwich Village and the Lower East Side, that he occupied over the next three decades. In the city, he soon found employment as a commercial artist but quickly learned that skilled commercial artwork paid much less than did house-painting (Lerner 1977, 334; Sylvester 1963, 45; Sandler 1989, 216). "It [America] turned out to be quite different from what I thought," he said. "Nowhere near as luxurious as

I imagined it" (Stevens and Swan 2004, 68). He returned to working as a house painter and lived in grinding poverty, finding little time to draw or paint. "The trouble with being poor," he later remarked, "is that it takes all your time" (T. Hess 1968, 101).

In Greenwich Village, de Kooning discovered a tradition of American painting and poetry that he had not known existed. He began visiting museums and commercial art galleries and became acquainted with the small underground community of émigré modern artists and intellectuals that congregated in the cold-water flats and cafeterias of Manhattan. Around 1929 he met John Graham, in 1931 Arshile Gorky, and somewhat later David Smith. Graham, Gorky, and Smith—"the three smartest guys on the scene"—all shared de Kooning's interest in modern art and were receptive to his view of the vital relevance of past masters (Rosenberg 1972, 54). Most important, they were mentors and role models who demonstrated that it was possible to be a modern artist in America. They were all intent museum and gallery visitors, studying and discussing the works in the Metropolitan Museum of Art, the Frick Collection, the galleries of Manhattan, and especially the Museum of Modern Art, which had opened in 1929. Over the years de Kooning saw countless works of art that profoundly influenced the course of his painting. New York during these years offered views of the work of Picasso, Matisse, Miró, the surrealist artists, Chaim Soutine, and Giacometti, as well as ancient art and works by the old masters. "You can safely say he saw everything," his future wife Elaine de Kooning later remembered (Yard 1986, 14).

Of his three companions, de Kooning readily acknowledged that Gorky, who exposed him to a broad range of art, was his most critical role model. Although the two artists were the same age, de Kooning called Gorky the "boss" (Mooradian 1971, 49). Both admired classical draftsmanship and the Western humanitarian tradition based on the figure. De Kooning recognized that although his schooling in Holland had been more rigorous than Gorky's in Russia, Gorky knew "lots more about painting and art—he just knew it by nature—things I was supposed to know and feel and understand—he really did it better" (Sylvester 1963, 46). He also acknowledged that Gorky's example helped him to develop the courage and conviction to become a serious painter. After Gorky's suicide in 1949, de Kooning wrote a letter to the editor of *ARTnews* affirming his debt to the late artist: "I am glad that it's about impossible to get away from his powerful influence" (De Kooning 1949a).

As soon as he could afford to buy supplies, de Kooning began to paint and draw on weekends, continuing to support himself during the week painting houses and doing carpentry. He also supplemented his income doing freelance commercial work, making signs, arranging window displays for department stores, composing designs for decorators, and painting decorative murals for speakeasies. He recalled these years as a time of financial struggle. He was self-conscious about his art and he kept secret his budding ambition to become a serious artist (Parsons 1967–68, 1). During the 1930s and, in fact, until the last half of the 1940s, he was constantly plagued by self-doubt. He finished few paintings and destroyed or painted over many others.

Becoming a Serious Artist

In 1935, in the middle of the Great Depression, de Kooning quit his job at Beck's Department Store, where he did most of his commercial design work, to work under the Federal Art Project of the Works Progress Administration (W.P.A.). He was hired by the W.P.A. for a little over a year, resigning when the relief program adopted a policy that required workers to be citizens. (De Kooning became an American citizen only in 1962.) He initially worked on the Easel Project, a division of the Federal Art Project; after six months he joined the Mural Division, where he worked on two projects.

Surviving sketches for one of the murals, as well as other abstractions from the last half of the 1930s, reveal, in their integration of biomorphic forms and geometry, the young de Kooning's debt to contemporary paintings by Gorky and European modernists such as Jean Arp, Piet Mondrian, Pablo Picasso, and Joan Miró. His work of this period also reflects his classical training at the Academy, where students were taught to precisely draw pure geometric forms.

De Kooning's work experience during his year on the W.P.A. significantly altered his self-image and impelled him to reverse his priorities and become a full-time artist. For the first time, he was able to devote himself entirely to his art, a decisive experience for him. As he explained to Irving Sandler:

> The Project was terribly important.... I changed my attitude to art. Instead of doing odd jobs and painting on the side, I painted and did odd jobs on the side. The situation was the same but I had a different attitude. I gave up the idea of first making a fortune and then painting in my old age (Sandler 1989, 216).

The Federal Art Project also put de Kooning in regular contact with new groups of modern artists, most of whom were in a similarly precarious financial situation. After work and on weekends, they would attend meetings at the Artists' Union or gather at cafeterias and studios for long discussions about art, artists, and current exhibitions. Despite the harshness of the Great Depression, de Kooning reported that there was a newfound sense of camaraderie among the artists. "The project was so good," he noted. "It brought us all together in a friendly way" (Parsons 1967–68, 1). Nonetheless, with no guaranteed income, he was still desperately poor and without any serious prospects as an artist.

A series of paintings of male figures executed from the mid-1930s through the early 1940s seem to respond to the mood of economic distress and melancholy that came from the artist's experience of the Depression. Often based on self-portraits (made from photographs and mirrors in his studio), a manikin he constructed, and studies of his friends and neighbors, such as poet and dance critic Edwin Denby and photographer Rudolph Burckhardt, they are anonymous, somber, and isolated. They are painted in deep, warm tones with indefinite outlines and combine allusions to art history, modernism, and everyday life. Prompted in part by Gorky's contemporaneous figure paintings, de Kooning later referred to the men as imaginary portraits in the manner of Ingres and the Le Nain brothers (Rodman 1961, 103).

De Kooning's art continued to develop in the late 1930s and early 1940s, shifting periodically from relatively abstract to more figurative modes, beginning an oscillation between these poles that would characterize his career as a whole. At the opening of the 1940s, he turned to the female figure for his subject matter, depicting women in beautiful and highly finished drawings. These works were sometimes commissioned as portraits and illustrations for fashion magazines. In 1937, de Kooning met Elaine Fried, whom he married in 1943. Elaine and her sister Marjorie may have served as models for some of these paintings (Zilczer 1993c, 24).

Like the paintings of male figures, de Kooning's women of this period combine references from traditional art history with modern interpretations of Hollywood starlets and fashion models. Unlike his depictions of Depression-era men, the women in most of these images are elegantly clad. The colors are more vivid and the fragmentation and dismemberment of forms is more radical in conception. Ingres's paintings at the Metropolitan, as well as works by Matisse and Picasso, were precedents for these paintings. De Kooning later recalled, "Picasso and Matisse showed us the way and we filled in with our own personalities" (Prather 1994, 80).

According to his acquaintances, de Kooning continued to be a harsh critic of his own work, full of doubt while working and unable to resolve the myriad available possibilities. As a result, most of his paintings remained unfinished for extended periods of time while he worked through countless scraped-away versions. However, by 1946, he began to execute abstract compositions in color and paintings, predominantly in black and white, that possess a vitality and formal inventiveness unprecedented in his earlier work. From this time onward, de Kooning was far more prolific than in previous years, destroying fewer paintings and finishing more. For some critics and historians, these forceful and complex abstractions set the boundaries by which his subsequent output was measured, remaining the unsurpassed achievement of the artist's career.[3]

De Kooning exhibited a group of these black-and-white paintings, including *Zurich* (fig. 16) and *Black Friday* at the Charles Egan Gallery in 1948, his first one-man exhibition. Although he did not sell a single painting and remained, according to Denby, "poorer than ever," this exhibition secured the first critical recognition for the forty-four-year-old artist and established him as one of the leading abstract painters of New York's avant-garde (Denby 1988, 13–14). A number of other solo and group shows held through the final years of the decade also focused on these essentially abstract images and received laudatory reviews. Critic Renée Arb, writing for *ARTnews*, wrote, "Here is virtuosity disguised by voluptuousness—the process of painting becomes the end" (Arb 1948, 33). Probably the most influential review of the Egan exhibition of 1948 was penned by critic Clement Greenberg. Writing for *The Nation*, he hailed de Kooning as "one of the four or five most important painters in the country" and as "an outright 'abstract' painter" who excluded recognizable imagery and referential content in his work to produce "an art that makes demands only on the optical imagination" (Greenberg, 1948b, 448).

During the 1940s, the loosely formed "downtown" modern artists, in general, worked with greater confidence and began receiving increased exposure in exhibitions and publications that sympathized with the avant-garde. Jackson Pollock, in particular, with the support of Peggy Guggenheim and her advisor Howard Putzel, began to achieve both critical and commercial success. The exhibition of his paintings along with those of the surrealists in 1943 at Guggenheim's Art of This Century gallery was a landmark in the emergence of modern art in America. Although de Kooning and Pollock were often viewed as competitors, there was also a sense of mutual admiration between the two artists. Pollock's method of dripping paint, in fact, may have strengthened de Kooning's interest in experimenting with unorthodox paint media applied in unconventional ways.[4]

Although the Abstract Expressionist artists themselves resisted a collective identity, they achieved a kind of solidarity as a self-proclaimed avant-garde movement between the end of World War II and 1950–51, and a certain consciousness about their place in American culture informed their activities and their conversations. On New Year's Eve of 1950, artist Philip Pavia announced to the party revelers, "The first half of the century belonged to Paris. The second half century will be ours." De Kooning added, "This is when it's going to happen," expressing the conviction, shared among this group of artists, that an advanced American art movement finally was receiving public exposure (Stevens and Swan 2004, 292). When the Metropolitan Museum of Art failed to include judges sympathetic to abstract art on the panel of its juried show *American Art Today, 1950*, the avant-garde artists galvanized and wrote an open letter of protest that was reported on the front page of the *New York Times*. That fall, *Life* magazine published an article on the controversial exhibition, along with a picture of fifteen of the eighteen excluded protesters under the headline "The Irascibles" (*Life* 1951).[5]

During this period, de Kooning began to attract wider interest among collectors and museums. *Painting* (1948), the largest work shown at the Egan Gallery show, was purchased by the Museum of Modern Art after it and *Excavation* (1950) were included in the Museum of Modern Art's 1951 exhibition *Abstract Painting and Sculpture in America. Excavation* was selected by Museum of Modern Art curator Alfred Barr—along with works by John Marin, Jackson Pollock, and Arshile Gorky—to represent the United States at the 1950 Venice Biennale. In the June issue of *ARTnews*, he called the selected painters a "predominant vanguard" (Barr 1950, 22). *Excavation* later won the Logan Medal and Purchase Prize and was subsequently acquired, amid considerable controversy, by the Art Institute of Chicago. The combined effect of this kind of publicity sent a powerful signal to the art world and the general public that the Abstract Expressionist painters were part of a significant movement, rather than a few radical artists.

With broadening exposure and greater notoriety, the group's social patterns began to change. During the 1930s and the early 1940s, the impoverished artists, largely ignored by the general public, had gathered in cafeterias and bars or had met in each other's studios to discuss matters of artistic, intellectual, and practical concern. By 1948, a number of them, wanting more privacy for their loose-knit artists' discussion groups than the cafeterias could provide, jointly rented a loft on East Eighth Street and dubbed it "The Club." Friday evening talks at The Club were initially given under the name "Subjects of the Artist" school. After the school was discontinued in 1949, the Friday talks continued under the name "Studio 35" (Sandler 1965, 27–31). Critics started to attend these sessions along with curators and collectors. De Kooning was not only a founding member of The Club, but by the end of the 1940s and into the early 1950s he also took part in several of the panels and symposia that member artists organized with increasing frequency. He later told interviewer Anne Parsons that the informal gatherings helped the New York artists out of their provincialism. "It was so important, getting together, arguing, thinking" (Parsons 1967–68, 3).

Achieving Critical Acclaim

From 1951 through 1954, de Kooning, now celebrated by artists and critics alike, suspended work on the abstractions that had earned him such success and returned to the figure. Although he had continued to work on the subject of the woman while he was executing his abstractions, few critics seem to have been aware of this fact. Throughout the rest of his career, he would turn with just such decisiveness again and again to new styles, resisting pressures to conform to the aesthetics prevalent among New York artists and critics. *Woman I*, the major painting of this series, was started in 1950 and went through many revisions before it was completed a year and a half later. While he worked on this painting, de Kooning produced other images of women in dozens of paintings and drawings (Carmean 1978, 154–82). The artist's new focus signaled an intensification of his expressionistic style with more painterly and violently distorted images. De Kooning himself said that the women provided a way out of an impasse: "Maybe I was stuck to a certain extent; I couldn't go on. It [the figure] did one thing for me: it eliminated composition, arrangement, relationships, light—all this silly talk about line, color, and form" (Sylvester 1963, 46).

When *Woman I* and five related paintings of women, along with several small oils on paper and pastels from the series, were shown in 1953 at the Sidney Janis Gallery in New York, they provoked intense discussion within the art community, fueling the ongoing debate between advocates of figuration and those who endorsed abstraction. Conservative critics complained about the artist's attempt to mix painterly abstraction and expressionist figuration. Champions of "advanced" art attacked de Kooning for his conservative return to figure painting and his retreat from the progressive imperatives of modern art. Although for two years Jackson Pollock had included the figure in his paintings, at the party following the exhibition opening he confronted de Kooning, whom he described as a "French painter," whose art still relied on recognizable imagery (Rodman 1961, 85). Several years earlier de Kooning had defended his refusal to devote himself exclusively to one kind of painting when the black-and-white abstraction *Painting* of 1948 was accompanied by the following statement:

> I'm not interested in "abstracting" or taking out or reducing painting to design, form, line, and color. I paint this way because I can keep putting more and more things in—drama, anger, pain, love, a figure, a horse, my ideas about space (Louchheim, 1951, 17).[6]

Critics in 1953 and commentators ever since, fascinated by the psychological charge of these works and by the artist's apparent hostile view of the female, have superimposed a multitude of personal and cultural meanings onto his depictions of women. As Shiff noted, "From its first public appearance, de Kooning's *Woman I* raised questions of the artist's personal attitude toward women" (Shiff 1994, 49). In hindsight, de Kooning acknowledged that the woman looked ferocious, but the images, rather than reflecting misogyny, were conceived as a satiric, comic homage to the ancient tradition of the female nude in Western art. "The Women," he claimed, "had to do with the female painted through all the ages, all those idols . . . Venus, the nude" (Sylvester 1963, 46). He also acknowledged contemporary sources: women seen on the street and in billboard and magazine advertisements (T. Hess 1968, 21). Elaine de Kooning said that these pictures were also concerned with the problems of figure painting: "They [the *Woman* paintings] do have a certain ferocity, but that has to do with paint" (Shiff 1994, 69). An abundance of commentary has followed in the wake of early critical responses to de Kooning's images of women, his own comments about the paintings, and renewed looks at the paintings themselves, making de Kooning's women among the most interpreted paintings of the twentieth century and undoubtedly his most well-known works among the public.[7]

After almost five years of working on the paintings of women, de Kooning again moved away from figuration. In 1955 and 1956, he painted a series of abstractions that Thomas Hess dubbed "abstract urban landscapes" (T. Hess 1968, 26). These brash and vital paintings evoke the robust sights, colors, and tempo of New York City. Paintings such as *Police Gazette* (1954–55) and *Gotham News* (1955) refer to the gritty squalor and bustling activity of Manhattan. The titles, sometimes supplied by friends, were frequently taken from the sex-and-violence tabloids and dime-store novels that often lay around de Kooning's studio (Cherry 1989, 230). A 1956 show of these works at the Sidney Janis Gallery was a tremendous success. The press praised the new work and the show sold out within a week of the opening.

The increasing critical acclaim that followed de Kooning's work after 1948 reached its apex during the 1950s. By the later part of the decade he at last attained a degree of financial success, emerging from the poverty that had plagued him since the Great Depression. For the first time, he could readily buy painting materials and was able to afford a large loft studio on Broadway. The 1950s also saw the emergence of a second generation of Abstract Expressionists, most of whom were deeply indebted to de Kooning for the development of their own gestural styles (Sandler 1978, 256–77).

De Kooning's fame increased in tandem with the growing recognition that the Abstract Expressionists were receiving in the United States and abroad. In 1956, *Time* magazine featured the work of these avant-garde artists in an article accompanied by several color photographs (*Time* 1956, 72–75). In 1958, the Museum of Modern Art organized *The New American Painting*, an exhibition that traveled to eight European cities and attracted large crowds. Clearly, the Abstract Expressionists were no longer a struggling underground. The group as a whole was becoming successful, celebrated, and sought after, and de Kooning was viewed as one of the leaders of the most advanced art movement in the New York art world.

The increasingly competitive and moneyed environment of the rapidly changing art world began to erode the solidarity of the group's working-class camaraderie. Personal pressures as well as those imposed by critics, dealers, and hangers-on were taking a toll on de Kooning, and he increasingly turned to alcohol as a way to escape his anxieties. His marriage to Elaine had collapsed and they separated in 1955. His personal life was now characterized by periods of binge drinking and an ongoing series of affairs. In January 1956, Joan Ward, a commercial artist who had had a relationship with de Kooning since the early 1950s, gave birth to his daughter, Johanna Liesbeth de Kooning.

In 1957, the congested energy of his abstract urban paintings gave way to the broad expanse of de Kooning's next series of paintings, his most abstract works dubbed "abstract parkway and pastoral landscapes" (T. Hess 1968, 26). Although de Kooning never learned to drive, he enjoyed making car trips between his studio in the city and suburbia along the parkways and expressways, studying the passing landscape. The paintings of this series suggest fleeting glimpses of the countryside viewed from a speeding car. Like the paintings of his good friend Franz Kline, these works convey the broad movement of the artist's arm and have come to epitomize "American action painting." Although they were executed in the city, they reflect the light, water, and color of Long Island.[8] An exhibition of these large new abstractions opened at the Sidney Janis Gallery in May 1959. By the end of the week, all of the paintings had sold.

As de Kooning continued to be lionized by his greatest supporters, his reputation in other circles was declining. In 1959, Thomas Hess published the first monograph on de Kooning. This influential book gave an intimate view of the artist. As critic Lawrence Alloway observed, much of its information was derived from studio talks and is "redolent of personal contact and transmitted opinion, straight from the horse's mouth" (Alloway 1975, 46). In 1960, critic David Sylvester conducted a now-famous interview with de Kooning that was excerpted and published as "Content Is a Glimpse" (Sylvester 1963). At the same time, rumors were circulating that de Kooning was washed up. His heavy drinking and indiscriminate socializing was becoming extremely self-destructive. In the two years following his successful 1959 exhibition of the parkway abstractions, he completed only a few full-scale paintings. Simultaneously, art fashion was shifting. The Museum of Modern Art's *Sixteen Americans* exhibition bypassed the older generation and emphasized the minimalist works of the young Frank Stella and Ellsworth Kelly. The 1961 show at the Guggenheim titled *The American Abstract Expressionists and Imagists* focused on the color-field abstractions of Morris Louis and Kenneth Noland. In 1962, the Sidney Janis Gallery opened a show called *The New Realists* that featured the work of leading pop artists Andy Warhol, Roy Lichtenstein, and James Rosenquist.

Keenly aware that a shift in contemporary taste was taking place and feeling a growing sense of isolation, de Kooning became increasingly disenchanted with the New York art world and began to dream of building a studio and moving permanently to Long Island. He later told an interviewer:

> New York is a great city, but in the long run I got kind of dried out by everything happening there day and night. I was less and less able to work regularly. Besides, I didn't really feel like running to all these artists' parties (Stevens and Swan 2004, 424).

Moving to the Country

In the winter of 1961 de Kooning began building a large studio-house on land he bought near his house in the Springs area of East Hampton. The studio was based on his own design, in collaboration with an architect, and was constructed with financial aid from collector Joseph Hirshhorn. "I'm tired of the rat race," he said, "but I need my friends around. Here I'm still within the circumference of New York, and I can look at the trees. Also, I want to live as well as the dealers" (Dickerson 1964b, 70). It was the beginning of his withdrawal from Manhattan. By the summer of 1963, his move from the city to the newly constructed studio was complete. Once established in Springs, he had no desire to paint anywhere else. He told Daniel Frasnay, "I am far removed from my Tenth Street studio and life here is so comfortable, I have no desire to go back to the city" (Frasnay 1969, 222).

De Kooning's first paintings after his move to Long Island were a renewed examination of the woman theme, but "cast in a new light" (Prather 1994, 174). In the paintings of this period, shimmering, voluptuous figures, usually indicated by strokes of puckered pink paint, were increasingly inseparable from the fluid paint medium, until by 1975 they approached complete abstraction. Although elements of the external world are less obviously evident in these paintings than in the preceding years, the local landscape, the water, and the light still play more significant roles. As Marla Prather has noted, "Like Monet at Giverny, he looked to his immediate world and his own art to create a series of monumental abstract landscapes that were among the most exuberant paintings of his career" (Prather 1994, 174).

In 1968, Thomas Hess organized a major retrospective of de Kooning's work for the Museum of Modern Art in New York. The exhibition opened at the Stedelijk Museum in Amsterdam and traveled to four other cities in Europe and the United States. A year later, de Kooning began making sculpture while continuing printmaking/lithography, which he had started in 1966. In 1975, he created a series of paintings in which figuration is increasingly subsumed by the painterly medium to the point of abstraction. In 1978, the Guggenheim Museum organized an exhibition that focused on de Kooning's paintings executed after his departure from New York (Waldman 1978). A 1983 exhibition at the Stedelijk Museum in Amsterdam also focused on de Kooning's later works, after his move to Long Island (De Wilde and Ratcliff 1983).

By the mid-1970s, de Kooning's health had deteriorated significantly as a consequence of his continued alcoholism. From 1978 to 1980, he produced so few paintings that his condition was threatening to end his career. At this point, Elaine, herself a recovered alcoholic, resumed contact with her husband and began to bring order to his life and work. As a result of her efforts, de Kooning slowly withdrew from alcohol and remained almost completely abstinent after 1980.

In the 1980s, de Kooning's last decade of activity, the densely worked surfaces and the chromatic intensity of the pictures from the late 1970s gave way to increasingly spare abstractions that have been referred to as the artist's "old-age style" (Rosenblum 1990). Works of this period have smooth surfaces, the result of the artist's use of scrapers to spread on and scrape off the paint. The palette is greatly simplified. Broad areas of creamy white and yellow fill the canvas and contrast with ribbonlike strokes of violet, pink, and blue that weave across the canvas surface. By the mid-1980s de Kooning had largely reduced his palette to the primary colors—red, yellow, and blue—laid down against a pure white ground. This decade was a period of enormously increased productivity.

In an exhibition that traveled in the United States and Europe in 1995, de Kooning's late work was presented to the public in a comprehensive manner for the first time; the artist, however, was unable to attend the exhibition, due to the deterioration of his memory, which had accelerated by the end of the 1980s. By this time, ill and housebound, de Kooning had ceased to paint, ending a career of almost six decades that had radically altered the shape of American art and that can be characterized by perpetual exploration, renewal, and rediscovery. He died in 1997.

De Kooning's Personal Statements and Interviews

Over the years, de Kooning's art and life have been thoroughly documented in monographs, reviews, and essays, in photographs and on film. Likewise, many interviewers prevailed upon the artist to explain his imagery and his work. In these published statements, de Kooning seldom constructed extended arguments to present his position. Instead he used homey, down-to-earth language and talked about his paintings and his working methods in short, pithy remarks, connecting disparate ideas by loose association. Likewise, he enjoyed informal give-and-take discussions about art with friends but was apprehensive whenever he was asked to record or deliver a formal statement. His apparently straightforward and spontaneous comments are rarely simple and can appear paradoxical. Nevertheless, because the artist often repeated sentences virtually verbatim from one statement or interview to the next, it is possible to discern certain recurring themes. These paraphrases help scholars evaluate the artist's attitudes toward art and give some insight into his technical methods.

In his first public statement, "A Desperate View," delivered in February 1949 at the Subjects of the Artist school, de Kooning stated, "In art, one idea is as good as another. . . . Art should not have to be a certain way. . . . Style is a fraud" (De Kooning 1949b). According to the artist, no single "idea" could define art. To rely on a predetermined artistic style was to close a painter's mind to new insights and possibilities. De Kooning repeatedly stated his opposition to all dictatorial systems that established notions of exclusiveness in art or prophesied the "progress" of art, whether they were rules and regulations designed to govern the production of a painting or ideologies in art that explained the development of its history. Although he greatly admired the work of Piet Mondrian, he said, "It was a horrible idea of van Doesburg and Mondrian to try to force a style. . . . I think it is the most bourgeois idea to think one can make a style beforehand" (De Kooning 1949b). De Kooning consistently avoided categorizing his art, just as he avoided constraining his work to a single artistic style or working method.

De Kooning's second and third public talks, "The Renaissance and Order," delivered at Studio 35 in 1950 (De Kooning 1951a), and "What Abstract Art Means to Me" of 1951 (De Kooning 1951b), given at the symposium organized by the Museum of Modern Art on occasion of the exhibition *Abstract Art in America*, expressed similar views. Both talks were given at a time when the artist had moved away from the black-and-white abstractions and was struggling with *Woman I* and his return to the figure. Each talk challenged prejudices developing in the art world and overtly criticized his contemporaries—artists and critics—who were, in his view, attempting to dictate the direction art should take. He was specifically distrustful of those theorists and visual artists who were turning "abstraction" into a program.

De Kooning often described himself as an eclectic artist, influenced by the diverse legacy of Western painting and sculpture (Rosenberg 1972, 54; Parsons 1967–68, 4). Modern art, he argued, was losing its touch with the visceral world and the human-centered approach of the Western tradition. He expressed admiration for the Mesopotamian idols and figures in the Pompeian frescoes, and he respected artists as disparate as Rembrandt and Cézanne, Mondrian and Picasso, Ingres and Soutine, who had taught him things that he could incorporate into his paintings. In particular, he celebrated the earthiness of the Renaissance artist:

> When I think of painting today, I find myself always thinking of that part that is connected to the Renaissance. It is the vulgarity and fleshy part of it which seems to make it particularly Western (de Kooning 1951a).[9]

De Kooning consistently appreciated works of art from the past and believed that its influence on his work was a healthy one. This is reflected in his remark, "I have no message. My paintings come from other paintings. . . . This is my pleasure of living—of discovering what I enjoy in paintings" (Rosenberg 1972, 57). Similarly, he did not hold to the idea, espoused by his colleagues Clyfford Still and Jackson Pollock, that it was possible to reject the past and create art anew.[10] He felt that as a modern artist he was part of a continuum, inspired by art of the past and, at the same time, inevitably adding to the course of art. "It is no use worrying about being related to something it is impossible not to be related to. . . . Whatever an artist's personal feelings are, as soon as an artist fills a certain area on the canvas or circumscribes it, he becomes historical. He acts from or upon other artists" (De Kooning 1949b).

Heavily influenced by other art, he was also greatly affected by the world around him, as mentioned above. More than most other Abstract Expressionists, de Kooning took inspiration not only from his surroundings and everyday environment but also from American popular culture: the bold graphic style of magazine ads, tabloid newspapers, Hollywood movies, store signs, and highway billboards. Thomas Hess elaborated on the significance of this imagery: "The concept of the American banal has always intrigued de Kooning: advertising, commercial art. . . . He found it romantic, vulgar, and in its own way poignant . . . and he decided he liked this intrusion of the art of the streets into his work" (T. Hess 1968, 76). During his youth and middle age, de Kooning was keenly aware of the sights and sounds of the streets of lower Manhattan, where he lived and frequently walked at night. In later years, he absorbed the atmosphere of eastern Long Island, bicycling along the flat potato fields, sitting on the Atlantic shore, and watching the changing light on the ocean.

These diverse experiences and images found their way into de Kooning's art. "Forms ought to have the emotion of a concrete experience," he said, acknowledging that when he worked, he tried to capture in his paintings the feeling of a place or the emotion evoked by his surroundings (Sylvester 1963, 47). De Kooning's work, however, was never illustrative. Reality and references to commonplace or inconsequential events were allusive, telescoped into shapes and colors to suggest fleeting images quickly viewed. In his search for fresh ways to render his subject matter, he evolved innovative technical methods and played with a range of materials, both traditional art supplies and retail industrial materials. The results were often startlingly original.

Critical Reception of de Kooning's Paintings

The public's conception of de Kooning's work was largely shaped by critics Thomas B. Hess and Harold Rosenberg. Hess, the editor of *ARTnews*, and Rosenberg, a critic and poet, were both close friends of the artist. They produced the major texts on de Kooning beginning in the 1950s and brought wide attention to his work. They also established the interpretive framework within which he was generally defined and discussed for the next twenty years. They constructed their readings of the artist's work and practices in the context of the ongoing debate over the nature of advanced art in the postwar period. They emphasized the quality of ambiguity and struggle in de Kooning's paintings and championed him as an existential artist-hero. In "De Kooning Paints a Picture," which appeared in the March 1953 issue of *ARTnews*, Hess chronicled the artist's eighteen-month struggle to create *Woman I* (Museum of Modern Art; T. Hess 1953).[11] Illustrated with six photographs by Rudolph Burckhardt of the painting in progress, the article set a precedent for extensive visual documentation of the artist's painting process. In the essay, Hess compared the self-doubting process of decision-making that went into creating the painting to a philosophical journey in which easy successes were disdained, one in which the method or means of production was as important as the final image.

Hess's article on *Woman I* appeared just two weeks before the painting, along with five other large oils and sixteen pastels and drawings of figurative works, were exhibited at the Sidney Janis Gallery. Just three issues prior to publishing Hess's article, *ARTnews* had printed Harold Rosenberg's "The American Action Painters" (Rosenberg 1952). This momentous article on the Abstract Expressionist movement covered a mix of political, social, and aesthetic issues and gained the featured artists immediate notoriety. According to Rosenberg, the new "American Action Painters" were giving direct expression to the alienation they felt in a debased society that did not understand their motivations. Rosenberg stressed the moral position of the self-doubting artist who was involved in the creative process through the very act of painting. Rather than a space in which to "reproduce, redesign, and analyze, or express an object," the canvas became "an arena in which to act." The "gesture on the canvas was a gesture of liberation" in which the artist was free to develop his individual vision, whether it be the depiction of the figure or an abstraction. Although no artist was named, Rosenberg's essay was based on de Kooning's example, a fact that Rosenberg later confirmed when he wrote, "De Kooning's improvisations provided the model for the concept of Action Painting" (Rosenberg 1963, 134). Due in large part to Rosenberg's popular characterization of the new movement, the "Action Painters" early on became the dominant wing of the Abstract Expressionist movement, and de Kooning came to be considered in many quarters as the quintessential Abstract Expressionist, especially after Pollock's death in 1956.

Throughout the 1950s and into the 1960s, de Kooning continued to receive ever-broader exposure through monographs, one-man shows, and large survey exhibitions of contemporary art, which secured his position at the center of the most advanced movements in the international art world. The 1968 exhibition mentioned earlier, which was organized by Hess as the first major retrospective of de Kooning's work, also traveled to London, New York, Chicago, and Los Angeles, after opening in Amsterdam. The catalogue to the exhibition documented the numerous changes of style and the sheer volume of de Kooning's work, dividing the paintings into categories by period and subject. Hess suggested that while the artist's paintings formed groups of closely related works during periods of concentrated activity, cyclical reassessment was also characteristic of his practice. Hess also vividly described some of de Kooning's working methods, based on his firsthand observations and discussions with the artist during studio visits (T. Hess 1968).

Although many of de Kooning's old supporters remained loyal, his increasing fame met with a growing number of dissenters. In particular, critic Clement Greenberg and his colleagues, who set out to establish criteria for advanced modern painting, were critical of de Kooning's work after his black-and-white abstractions. In a prominent essay entitled "'American-Type' Painting," written in 1955 for *Partisan Review*, Greenberg dismissed de Kooning as a "late cubist," arguing that for all his talents, he was trapped by the past and was not making original American art. He added, "If de Kooning's art has found a readier acceptance than most other forms of abstract expressionism, it is because his need to include the past as well as forestall the future reassures most of us" (Greenberg 1955, 183).

By the late 1950s, a number of critics were beginning to question, in particular, the tenets of "action painting," claiming that the gesturalism of de Kooning's emulators had degenerated into clichés and was turning the unpremeditated spontaneity and immediacy of their brushwork into a mannered style.[12] Among the younger artists, de Kooning, so long viewed as a renegade, came to symbolize the ultimate insider. As critic Mark Stevens noted: "He became the Salon painter against whom Pop, Minimalist, and Color-field artists defined themselves" (Stevens 1994, 28). This view prevailed among critics and artists for many years. During the 1960s, 1970s, and early 1980s, de Kooning was labeled an old-fashioned painter who had made important work earlier in his career but who had fallen into decline and no longer made "advanced" art.

Over the past several decades, a new generation of scholars has reevaluated and revised Hess and Rosenberg's interpretations and the general perception that de Kooning's paintings after about 1960 were inferior to his work of the 1940s and 1950s. A series of monographs (Waldman 1988; Wolfe 1996; Yard 1997 and 2007; B. Hess 2004) and a recent biography (Stevens and Swan 2004) provide new insight into the artist's development over six decades and give evidence of de Kooning's continuing reputation and the need for both broader and more detailed information about his life and career. Additionally, a number of recent museum exhibitions have focused both on different periods of his painting and on the full body of his work, and they have provided greater evidence of the diversity of de Kooning's art within the context of his overall career.[13]

Literature on de Kooning's Materials and Working Methods

Although the literature on de Kooning is voluminous, few publications have dealt with his materials and working methods. Most references to de Kooning's painting practices and his materials are casual comments in art historical texts or interviews. In his many essays and monographs, critic Thomas Hess made frequent mention of the artist's skillful and highly idiosyncratic use of art materials. He wrote about de Kooning's use of charcoal and industrial enamel paint, and particularly on his use of drawings as a parallel activity to his painting. He also described the artist's practice of scraping down some or all of a painting, his preference for hard, smooth surfaces on which to paint, and his use of a fluid, slick painting medium that he associated with freshness.[14] Hess also described in some detail the artist's practice, during the 1960s and 1970s, of mixing artist's oil paints with safflower cooking oil, water, and one or more solvents (T. Hess 1978, 236–37).

More recently, a few scholars have expanded on Hess's observations. Sally Yard's 1980 dissertation on de Kooning's early years in New York included interviews with both the artist and his wife Elaine, which contradicted the widely held perception that de Kooning's paintings were unpremeditated. Instead, Elaine discussed the calculation involved in de Kooning's conception of his paintings, despite their appearance of gestural immediacy (Yard 1986). Harry Gaugh's "Notes on Technique" section of his book on de Kooning briefly summarizes many of Hess's observations (Gaugh 1983). A 1993 essay, to which the current author contributed, used infrared photography and reflectography to show that de Kooning throughout his career relied on preliminary drawings to compose and organize many of his compositions (Lake and Zilczer 1993). Two essays, by Marla Prather and Richard Shiff, in the catalogue for the 1994 National Gallery of Art exhibition of de Kooning's paintings include valuable insights into de Kooning's painting methods provided by the artist's friends and colleagues (Prather 1994; Shiff 1994). Klaus Kertess's essay "Willem de Kooning in Process" discussed the artist's working methods from 1970 to 1987, many which were extensions of his practices since the 1940s (Kertess 2000). While these studies added valuable insights into the artist's stylistic development, none of the discussions were corroborated by technical examination of the works themselves. In general, scholarly attention has focused on the historical development of Abstract Expressionism and on its subject matter. No in-depth examination of the physical appearance of the pictures and how they were painted has hitherto been undertaken.

Although the general character of de Kooning's methods and his materials has been described, writers have tended to reiterate Hess's observations verbatim. Scholars, for example, regularly echo Hess's statement that de Kooning's use of black enamel paint in his black-and-white paintings of the late 1940s is the only instance in which he used retail trade paints. Likewise, they repeat Hess's assertion that, with this exception, de Kooning prior to about 1960 used conventional artist's oil paints, simply applying them in unconventional ways (T. Hess 1953, 65).

Hess's descriptions of de Kooning's painting techniques have also been applied retroactively and universally. The artist's use of slick, smooth supports and his partiality to fluid paints, for example, are often discussed as if they persisted throughout his career (Shiff 1994, 53–54). In one case, a conservator, knowing that de Kooning used safflower oil later in his career, speculated that problems of condition related to an earlier painting by the artist might have resulted from safflower oil (Tomkins 1987, 66).[15] Moreover, Hess's description of de Kooning's mixture of safflower oil and water seems to have spawned rumors that the artist incorporated mayonnaise into his medium (Waldman 1978, 26; Russell 1978, sec. 2, 25; Anfam 1990, 180). These accounts might be dismissed as fanciful were it not for the fact that a 1978 Metropolitan Museum of Art Interdepartmental memo from John Brealey, chairman of the Department of Paintings Conservation, to Thomas Hess, then chair of the Twentieth-Century Art Department, mentions their discussion of different paint media during de Kooning's visit to the museum. The memo includes a recipe for an egg, water, and oil binding medium, "if Mr. De Kooning cares to experiment a little with this technique." There is no evidence that Hess conveyed the recipe to the artist.[16]

The current technical study not only considers broad issues of stylistic development, but also clarifies many of these areas of confusion about de Kooning's painting materials and practices.

2 Paintings of the Late 1940s and 1950s

The 1940s were years of daring experimentation for Willem de Kooning. During this decade, he established his own artistic style and subject matter, and by 1948 he had finally begun to achieve public, if not financial, success. As he shifted periodically from relatively abstract to more figurative modes, his style evolved from paintings with uniform surfaces to works in which he utilized more expressive brushwork and incorporated vigorous drawing methods. Significantly, he also used a range of different materials to evoke the situations and circumstances of his surroundings.

During an unpublished interview conducted in his studio in 1959 in the presence of several friends, de Kooning called himself a "slipping glimpser" and talked at length about the importance for his art of glimpsing tiny things, adding that when he worked he was not painting what he saw but was "illustrating the emotion of a concrete experience" ("Inner Monologue" 1959, 3–8). When critic David Sylvester interviewed de Kooning less than a year later, the artist spoke again about flashes or slipping glimpses, stating: "Content is a glimpse of something, an encounter like a flash. It's very tiny—very tiny content" (Sylvester 1963, 47). In these extended statements, de Kooning was acknowledging the importance to his art of his daily environment and the significance of random visual observations. He was also equating a painting's content or meaning with the emotion or experience of these brief, commonplace events. As Richard Shiff (1994, 53) observed: "The artist associated the glimpse with the most ordinary kind of 'happening'—somebody sitting on a chair, or a puddle of water reflecting light. He opposed it to things people think are important, 'whatever happens in newspapers.'" Thus, de Kooning's art throughout his career remained firmly rooted in his daily experience, and his keenly observed "slipping glimpses" are a source of much of the imagery in his paintings. Each painting is a node of memory, a record of commonplace or inconsequential sensations from his daily life, recorded after the event in the studio. As de Kooning noted, "If I really think about it, it will come out in the painting" (Sylvester 1963, 48).

In his interview with Sylvester, de Kooning also asserted that he was not an American but a New Yorker, acknowledging the significance of the city in his life and work during this period (Sylvester 1963, 46). Examination and technical analysis of de Kooning's paintings of the late '40s and '50s, have revealed that de Kooning imbued his art with an immediate sense of the urban environment in part by using unorthodox methods of paint application and by selecting and preparing paints from a variety of retail trade products.[17]

This chapter draws on the in-depth technical examination of seven paintings executed by de Kooning between the years 1943 and 1954. The paintings are *Queen of Hearts* (1943–46); *Special Delivery*, (1946); *Zurich* (1947); *Secretary* (1948); *Woman* (1948);

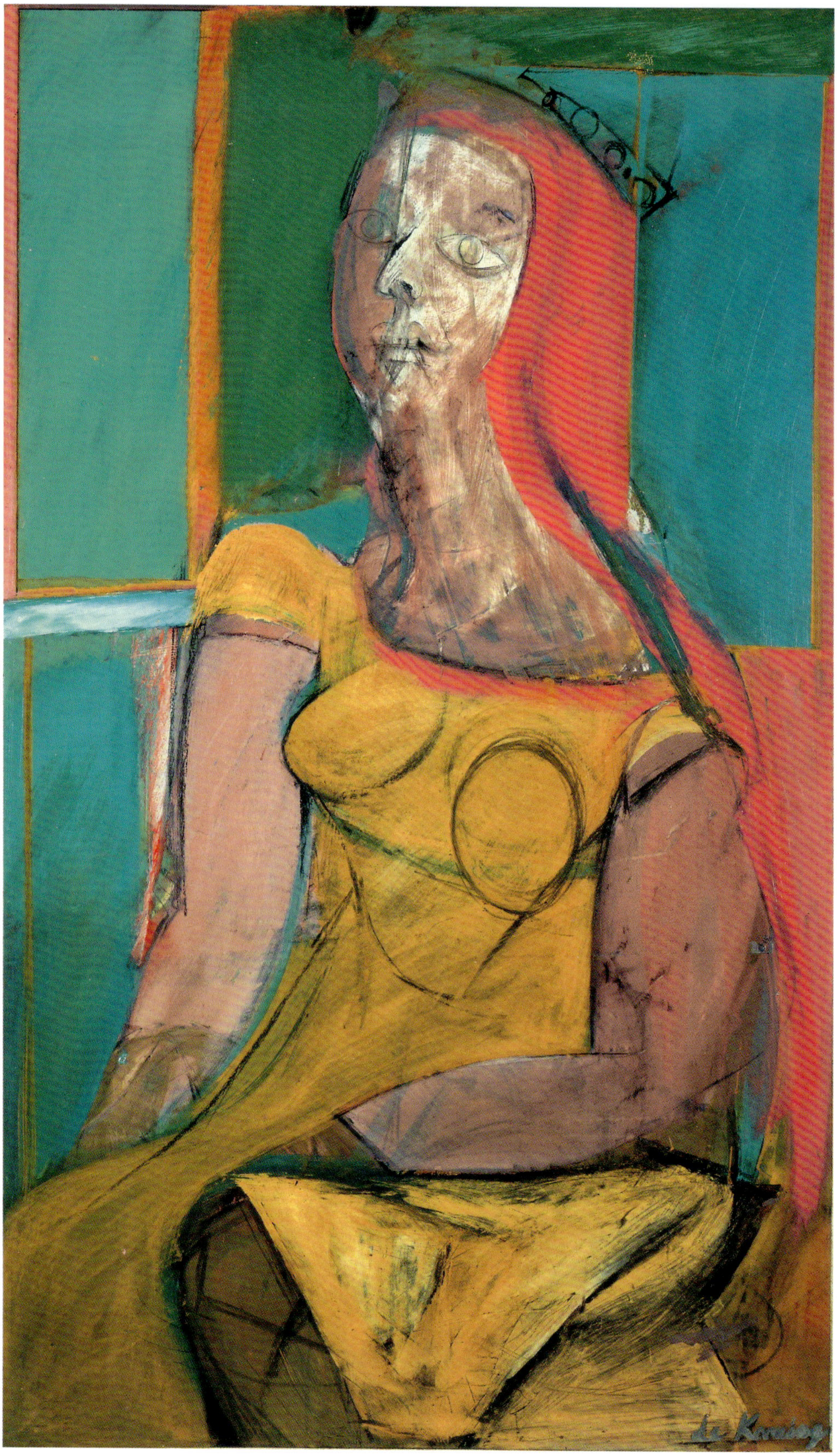

FIGURE 1
Queen of Hearts, 1943–46. Oil and charcoal on fiberboard, 117 × 70 cm (46⅛ × 27⅝ in.). Hirshhorn Museum and Sculpture Garden, Smithsonian Institution (66.1223). Gift of the Joseph H. Hirshhorn Foundation, 1966.

© 2010 The Willem de Kooning Foundation/Artists Rights Society, New York. Photo: Lee Stalsworth

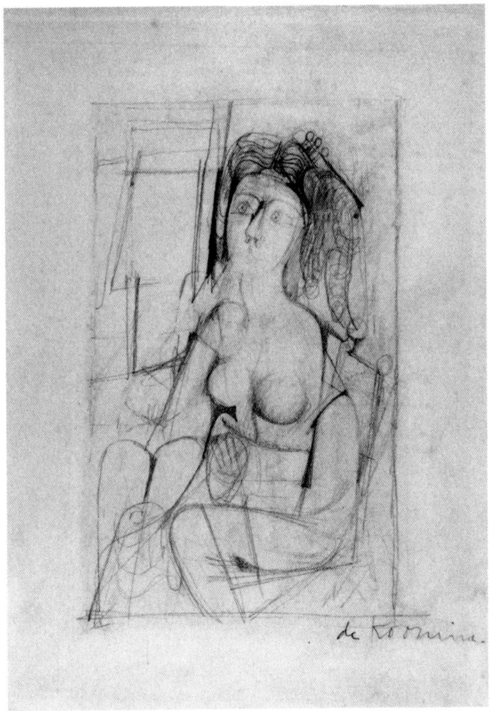

FIGURE 2
Seated Woman, 1943. Graphite on wove paper, 33.7 × 26 cm (13¼ × 10¼ in.). Philadelphia Museum of Art (2003.149.4a,b). Bequest of Elizabeth M. Petrie, 2003.

© 2010 The Willem de Kooning Foundation/Artists Rights Society, New York. Photo: Lynn Rosenthal

Woman (1953); and *Two Women in the Country* (1954). It also includes pigment analysis of *Valentine* (1947) and *Untitled* (1947–48), and medium analysis of *Black Friday* (1948), *Untitled* (1949), *Woman VI* (1953), and *Gotham News* (1955).

Early Representations of the Female Figure: *Queen of Hearts* (1943–46)

De Kooning worked in both abstract and figurative modes during the 1940s. *Queen of Hearts* (fig. 1), painted over a three-year period from 1943 to 1946, is an early representative of the female subject. It is also an important transitional work that retains traditional associations and also heralds his signature "women" of the late 1940s and the 1950s. Examination and technical analysis of the painting reveals that it evolved through a series of clearly discernible stages during its creation. It provides evidence of the artist's curious amalgam of traditional picture-making techniques and his use of less conventional painting materials and practice that later came to characterize much of his Abstract Expressionist work.

Queen of Hearts is rich with associations from the long history of this subject in art, which has been treated by artists from old masters to Ingres, Matisse, and Picasso. It also represents one of de Kooning's earliest attempts to reinterpret the traditional female figure in a relatively abstract mode. From 1940 through 1946, de Kooning usually framed his images of women within colored geometric shapes that allude to windows, pictures, or furniture. In this painting, the three-quarter-length subject sits close to the picture frame in a shallow, spatially unclear setting demarcated by intensely colored rectangular shapes. Seated with one knee raised, she is clad in a bright yellow dress and wears a pink headpiece and a crown.

Although the painting powerfully suggests a process of spontaneous execution, de Kooning, in fact, closely followed traditional practices in his use of preliminary drawings, his preparation of the support, and his initial application of paint layers. Unlike most of the Abstract Expressionists, de Kooning produced numerous studies and sketches as he planned and composed his paintings. While his use of drawings would in time develop into a working method based on a complex and symbiotic relationship between drawing and painting, in the early stages of composing *Queen of Hearts* the artist still adhered to traditional methods: he produced preparatory drawings that were worked out in advance or during the painting's execution and were then transferred to the painting support.

Seated Woman (1943; fig. 2), a graphite drawing on paper, served as a study for *Queen of Hearts*. The disjointed fragmentation of the three-quarter-length seated figure, evident in the painting, was conceived first in this study. Although the paint layers in *Queen of Hearts* are too thick for much of the preliminary drawing to be fully visible by infrared photography, brushy lines around the figure and in the area of the woman's head confirm the degree to which the underdrawing is faithful to the study. De Kooning probably used ink or a fluid black paint to sketch in the basic components of the figure, transferring the major elements of the pencil drawing to the white ground of the painting. He also chose to eliminate the woman's hair and add an ear, which is still discernible beneath the paint of the pink veil. The drawing also contains a heartlike shape in the lower left corner; infrared photography reveals that a comparable shape was once included in the painting but that the artist subsequently painted it out.

De Kooning's preparation of his supports also reflects a continuing reliance upon traditional methods at this stage in his career. To the smooth side of a sheet of Masonite, he applied several layers of a traditional gesso, made up of ground gypsum ($CaSO_4$) mixed with a water-soluble adhesive, probably hide glue.[18] Cross-sections of paint show that once the ground was applied and sanded to a smooth surface, de Kooning sketched in the ink underdrawing and then coated it with a thin layer of resin, probably shellac.[19] Conrad Fried, Elaine's brother and a painter who frequently worked in de Kooning's studio during this period, recalls the artist's careful preparation of his supports and remembers that de Kooning often added a layer of shellac to the priming layer of his paintings to decrease the absorbency of the ground (Prather 1994, 82). The use of a clear resin layer such as shellac over a preliminary drawing is a prescribed studio practice that de Kooning would have learned as a student and one he surely borrowed from his commercial experience. When designing an advertisement or a display in the 1930s and early 1940s, de Kooning would apply a coat of shellac over pencil drawings executed on cardboard supports primed with a gesso ground. Then he would apply printer's ink to the resin-coated surface. This preparatory method, now applied to his works of art, allowed him to change the painting as he worked, wiping or scraping the paint from the shellacked surface without damaging the pencil drawing beneath the shellac (Shiff 1994, 35).

Although *Queen of Hearts*, painted over a three-year period, shows little evidence of the repeated scraping and reworking so characteristic of many of his later paintings, de Kooning arrived at the final color scheme only after significant revisions. Over the preliminary drawing and layer of shellac, he blocked in the colored rectangular shapes of the background using a succession of different colored paints applied around the reserved area of the figure. Though the background zones of the picture now present the appearance of a uniformly smooth, thinly painted surface, de Kooning in fact created the current image through numerous alterations in color. A paint cross-section from the upper left corner reveals four distinct paint layers over the white ground (fig. 3). De Kooning first rendered this background area in a thick layer of cadmium yellow, then in cadmium orange, followed by a layer of pink, and then a layer of blue-green paint.

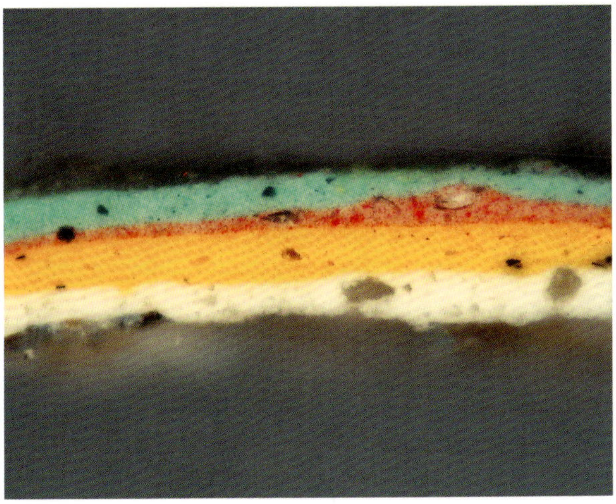

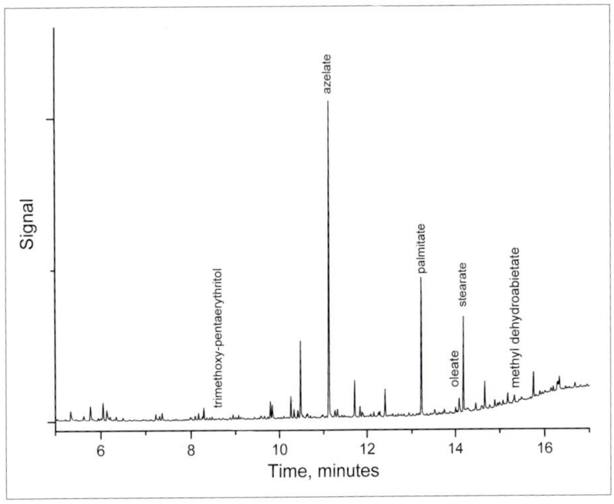

FIGURE 3
Paint cross-section from the top left corner of *Queen of Hearts*, showing four distinct paint layers over the white ground. Two additional layers are missing: masking tape and the pink of the final image (250x magnification).

FIGURE 4
Pyrogram (analysis by pyrolysis-GC/MS, see Appendix 2), *Queen of Hearts*, sample of the orange paint from the figure's proper right shoulder.

A strip of masking tape that the artist applied to the edge of the panel face and the pink of the final image are missing in this cross-section. From this sample and others taken from the painting, it is evident that the artist worked on the background over an extended period of time, building up the paint layers in distinct stages. The fact that the upper paint layers do not disrupt the brushstrokes of the earlier layers indicates that sufficient time elapsed between successive repainting for each paint layer to dry before the next was applied.[20]

Until the mid-1940s, the surfaces of de Kooning's paintings tended to be very smooth and somewhat glossy. His interest in creating uniform, shiny surfaces reflects the influence of Jean-Auguste-Dominique Ingres. After seeing the Metropolitan Museum's newly acquired paintings by the French painter, de Kooning, Gorky, and Graham adopted the master's smooth surfaces as well as his linear rendering of the figure. Friends of de Kooning and Gorky later described these artists' methods of carefully polishing their painting surfaces with a razor blade or with sandpaper to approximate Ingres's glossy surfaces (Lader 1978).

De Kooning's choice of paints was also important in achieving the porcelain-like surfaces he sought. The relative amounts of saturated palmitic and stearic acids, plus the presence of oleic (mono-unsaturated C_{18}) acid, and azelaic acid (the degradation dicarboxylic C_9 compound) indicate that the binding medium of these paints is a drying oil, probably linseed oil (fig. 4).[21] The identification of the medium as linseed oil, in conjunction with pigment analysis, supports the conjecture that the paints in this picture are artists' tube paints. The yellow, pink, and orange paints all consist predominantly of the relatively expensive cadmium pigments, with none of the resins and only small amounts of the fillers and extenders commonly found in house or sign paints of the period.[22]

Technical examination of *Queen of Hearts* also reveals that de Kooning was relying on his technical training, specifically his experience with and admiration for the craft of sign painting, to achieve his glossy surfaces.[23] Fine, circular striations in the green and blue rectangles behind the seated woman indicate that de Kooning enhanced the smooth surfaces of the oil paint by rubbing them with pumice stone or fine-grit sandpaper. He then applied a natural resin to the sanded surface before he signed the painting, to create paint films that appear uniformly glossy.[24] De Kooning's careful preparation of his paint surfaces is an additional instance of his adapting commercial methods to his studio practice. When traditional sign painters wanted to achieve a highly polished background on which to apply their lettering, they generally rubbed down their paints with lump pumice and water to remove all brushstrokes, and then gave the sign a coat of clear or tinted varnish. This process could be repeated several times (International Correspondence Schools 1899).

Microscopic examination and x-radiographs of *Queen of Hearts* indicate that de Kooning painted the seated figure only after he had completed the background.[25] This area gives evidence of an approach to working very different from the slow buildup of distinct paint layers evident in the construction of the background. Here the conventional techniques of painting were ignored for a more immediate handling of the materials, as the artist appears to have quickly and spontaneously assembled the segmented, unmodeled blocks of the woman's figure in one, at most two, layers of paint. It is as if he were following the advice of his friend Gorky to "keep the forms flat while retaining a feeling of physical bulk" (Yard 1986, 13).

De Kooning rendered the disconnected arms and head of the woman with the broad, sharp blade of a tool, such as a metal scraper. With this tool he first applied a layer of white lead paint directly over areas of the white ground, overlapping the edges of the surrounding green and blue background. As he worked on the woman's head, he pivoted the blade to form her profile and weak chin. When he reached the neck, he lifted and reset the scraper several times as he proceeded down the surface of the painting, allowing the tool's sharp edge to make a series of horizontal gouges into the wet paint, which look curiously like the wrinkle lines of sagging skin. Once the white paint dried, the artist next applied a thin wash of pink paint with a brush; this layer he, in turn, scraped down so that the lower white layer shines through the pink to give the woman's skin its translucent quality.

De Kooning's rendering of the figure's yellow dress and arms is also a radical rejection of established painting procedure. Examination with the microscope reveals none of the careful build-up of distinct paint layers that is characteristic of the background. Instead, it shows the degree to which the underlying charcoal fibers from a dense lower scribble in charcoal were worked up into the dilute wash of yellow paint when it was applied. The profuse charcoal lines that partially contour the woman's various body parts and scrawl across the paint surfaces are an element of crucial importance in de Kooning's paintings of this period. Aptly described by Sally Yard as "willful pentimenti," they have a very different function from the largely invisible ink underdrawing that served as a preliminary guide to the painting (Yard 1986, 104).

These multiple contours in charcoal suggest alternative positions for eyes, legs, shoulders, arms, and breasts. As such, they initially imply that de Kooning's working method involved early but still visible revisions to his composition, with each body part drawn in charcoal, painted, redrawn and repainted several times. In fact, close examination of the picture surface reveals that all of the charcoal lines lie directly on the paint surface or just below a translucent top paint layer. The woman's faint upper set of eyes, located above the almond-shapes ones, for example, initially appear to be an abandoned set submerged below an upper layer of paint. These "original" eyes bear a close resemblance in shape and position to those in the ink drawing *Head of a Woman* (*Study for Queen of Hearts*) (fig. 5), a study done while the painting was in progress. It is therefore tempting to postulate that de Kooning created this face at an early stage of the work and then painted out the eyes, which are now partially visible as pentimenti. In fact, both sets of eyes—the upper wide-eyed set in charcoal and the lower ones in graphite—are inscribed directly into the upper paint layer; the upper charcoal ones are simply smudged to reduce their prominence (fig. 6). In a similar manner, the multiple ovals that form the woman's breasts lie directly on the paint surface or just below a thin wash of paint. As he worked, de Kooning moved the breasts' position downward, retaining evidence of the earlier versions.

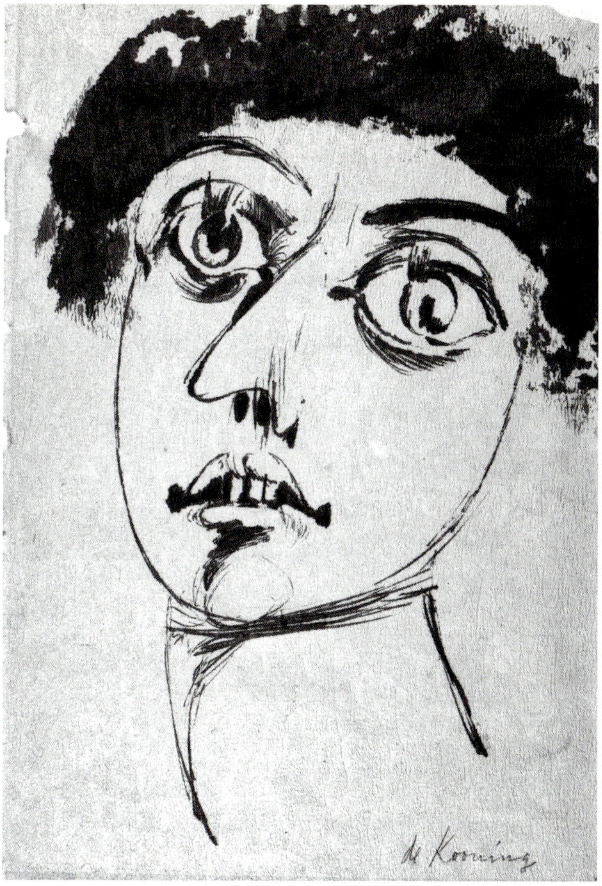

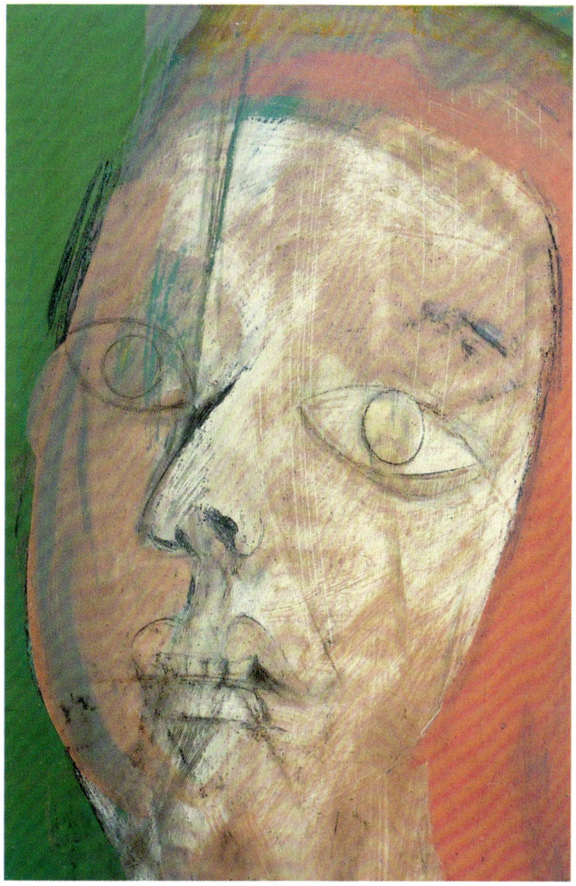

FIGURE 5
Head of a Woman (Study for Queen of Hearts), ca. 1943–45. Pen and wash on brown wove paper (paper towel), 28 × 20 cm (11 × 7⅞ in.). Iris & B. Gerald Cantor Center for Visual Arts at Stanford University (1984.224). Gift of Victor and Paula Zurcher.

© 2010 The Willem de Kooning Foundation / Artists Rights Society, New York

FIGURE 6
Detail of *Queen of Hearts*, showing the artist's use of charcoal and a scraper to apply and manipulate his paint.

Clearly, de Kooning did not attempt to conceal his revisions in charcoal. To achieve a compositional result, he intentionally allowed his alterations to show, preserving previous versions of his work. The lines in charcoal and the roughly brushed and scraped paints were deliberate devices to convey the immediate expressive gesture, the physical process that brought the work into being. As such, they herald the direction of modern and specifically Abstract Expressionist emphasis on the picture surface, manifesting the artistic work that goes into the construction of the picture and deemphasizing the representation of appearances and the traditional rules and conventions of picture-making.[26] This remained de Kooning's practice throughout his career.

During the late 1940s and early 1950s, the question of when a painting was finished or whether it was ever finished preoccupied de Kooning. His difficulty in finishing his paintings was legendary, and his method of resolving the image by working and reworking it soon established his reputation among his peers as an artist who never completed a painting.[27] According to Elaine de Kooning, the difficulty lay, in part, with the fact that no one bought contemporary American paintings. Unsold or abandoned paintings lying around the studio "became their own justification" (quoted in Janis and Blesh 1960, 25). De Kooning's friend Edwin Denby remembered that the artist was trying to master fundamental painting problems, such as the figure–ground relationship. He often heard de Kooning say that "he was beating his brains out about connecting a figure and a background" (Denby 1964, 87–88).

A detailed analysis of *Queen of Hearts* reveals just how deliberately unconventional de Kooning's approach was. Elements of traditional practice remain, for example, in the artist's carefully prepared ground layer, his use of a sketched study transferred

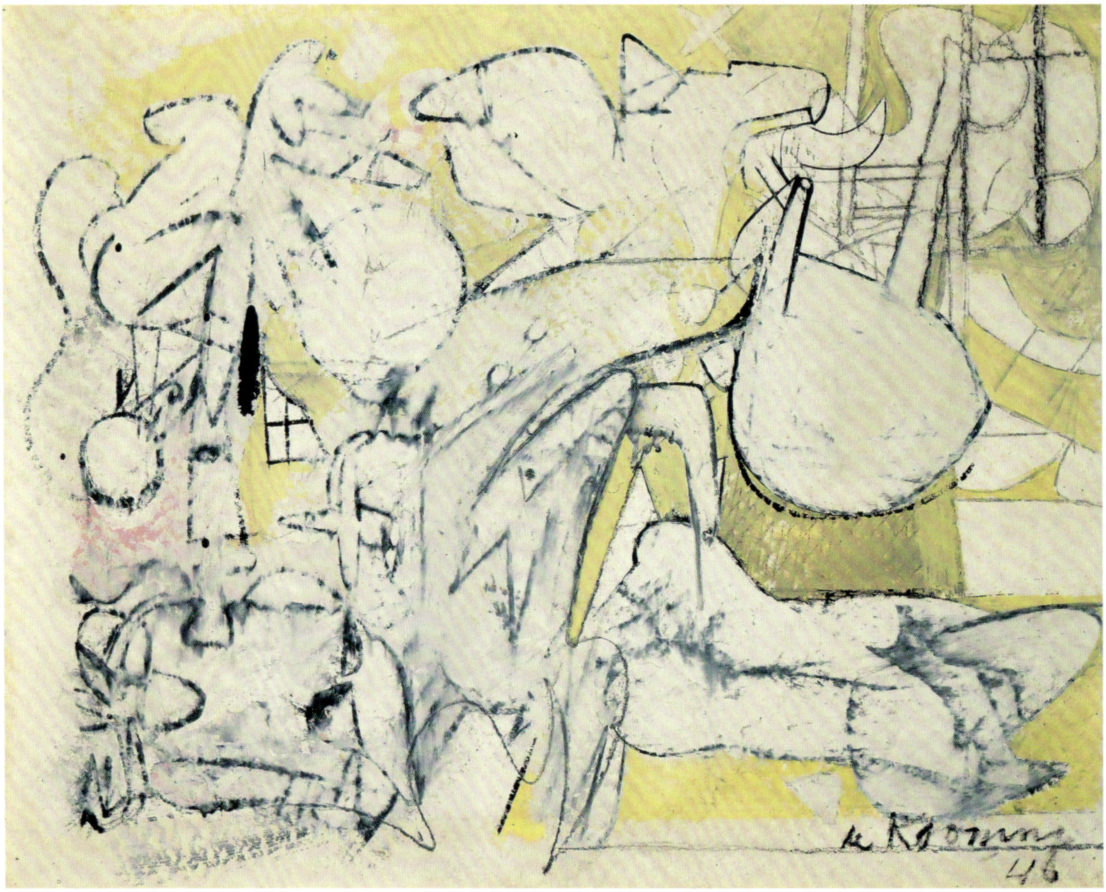

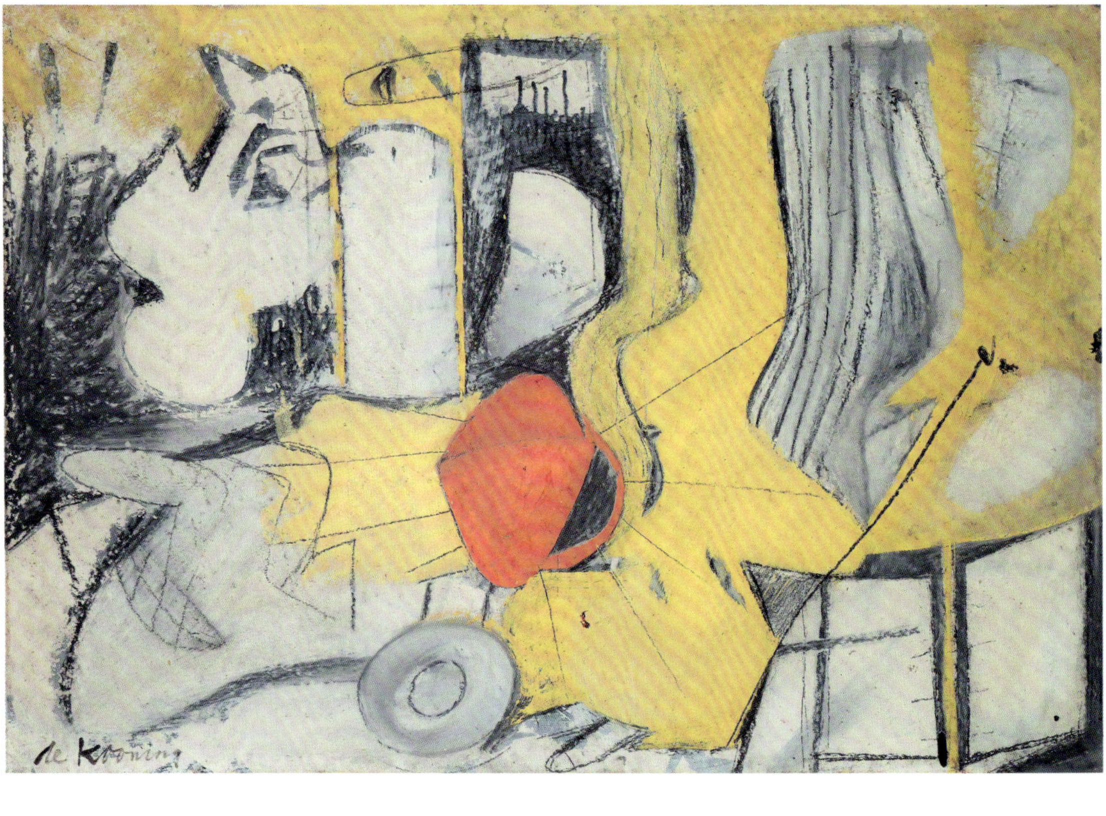

FIGURE 7
Special Delivery, 1946.
Oil and charcoal on paper mounted on fiberboard, 59.3 × 76.1 cm (23⅛ × 30 in.). Hirshhorn Museum and Sculpture Garden, Smithsonian Institution (66.1198). Gift of the Joseph H. Hirshhorn Foundation, 1966.

© 2010 The Willem de Kooning Foundation / Artists Rights Society, New York. Photo: Lee Stalsworth

FIGURE 8
Secretary, 1948. Oil and charcoal on paper mounted on fiberboard, 62 × 92 cm (24⅜ × 36³⁄₁₆ in.). Hirshhorn Museum and Sculpture Garden, Smithsonian Institution (66.1193). Gift of Joseph H. Hirshhorn, 1966.

© 2010 The Willem de Kooning Foundation / Artists Rights Society, New York. Photo: Lee Stalsworth

directly to the painting, and his careful buildup of successive paint layers, but overall the painting is remarkable for its breadth of handling. The lack of modeling or of any spatial recession and the clearly visible artistic revisions seem deliberately artless. Similarly, the combination of scraping and scumbled brushwork and the unconventional use of charcoal, in combination with highly polished passages, further undermine this work's unity. While it was clearly the result of an extended process of construction and experimentation, *Queen of Hearts* retains a sense of immediacy and even of improvisation that points ahead to *Woman* of 1948.

Colored Abstractions: *Special Delivery* (1946) and *Secretary* (1948)

From the mid-1940s on, the surfaces of de Kooning's paintings gradually evolved from the smoothness characteristic of paintings such as *Queen of Hearts* toward the more rugged impasto characteristic of the paintings of the late 1940s through the 1950s. Despite their small sizes, *Special Delivery* (fig. 7), painted in 1946, and *Secretary* (fig. 8), painted in 1948, are coarsely textured paintings executed in a limited range of colors that overlap with black paint and charcoal. In each work, de Kooning's "messy" methods of application appear deliberately crude in their handling. The artist used artistic means that demonstrate an extraordinary synthesis of accident and control, and he employed a variety of art and retail trade products whose properties he deliberately enhanced or manipulated in order to produce paintings reflective of the grimy urban environment of New York City. As his imagery became more abstract, de Kooning manipulated his materials in order to produce works whose textured surfaces evoke rather than illustrate urban life.

Although both *Special Delivery* and *Secretary* at first appear nearly abstract, they abound with recognizable imagery. The spatial configuration of *Special Delivery* suggests a room with a window and a ladder populated by teeming biomorphic shapes, some suggesting human and animal body fragments. The modeled, bright-orange form in the center of *Secretary*, presumably the secretary of the title, is connected by radiating diagonals to a caricatured face with a comic-strip grin, an opened stenographer's notebook in the lower right corner, and broadly suggestive curving and geometric shapes that resemble doors, legs, and breasts.

De Kooning appears to have achieved the amorphous, biomorphic shapes and incongruous space in *Special Delivery* and *Secretary* through a working method that involved superimposing his drawings and sketches to create composite images that were both randomly and deliberately formed. By layering his drawings—or torn fragments of them—one on top of the other and tracing new shapes from the composite forms, he could continually discover unforeseen new images that were eventually incorporated into his paintings (T. Hess 1968, 47). According to Thomas Hess, the imagery of many of the abstract paintings of this period evolved from drawings of the female figure that de Kooning had cut apart, transposed, and otherwise rearranged on the painting surface until they became "abstract," but they always retained a likeness of their source. He apparently began making paintings from composites of individual parts of his drawings as early as 1943 (Stuckey 1980, 74). A photograph taken in 1946 by Harry Bowden illustrates de Kooning's method (fig. 9). It shows a drawing of a woman's torso taped to an abstract painting in progress. De Kooning probably traced the drawing, and others scattered on the floor, from one of the adjacent figure paintings and ultimately produced the abstraction from an elaborate process that involved disassembly of the figure.[28]

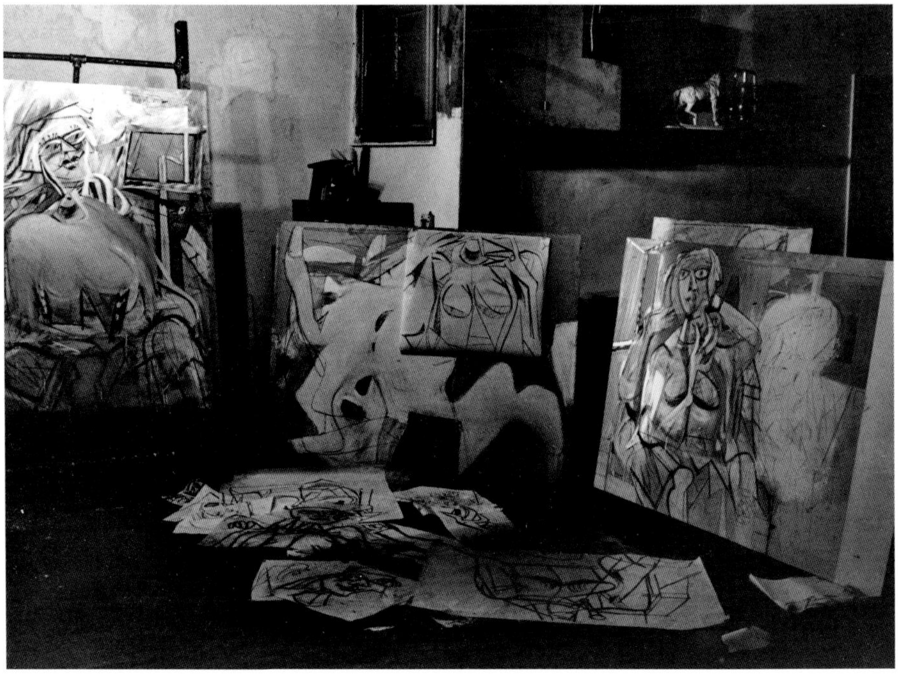

FIGURE 9
The interior of de Kooning's studio, 1946 (detail). Harry Bowden. Gelatin silver print, copy print, 19.1 × 19.1 cm (7½ × 7½ in.). Iris & B. Gerald Cantor Center for Visual Arts at Stanford University (1979.100). Gift of Charles Campbell.

Artwork © 2010 The Willem de Kooning Foundation / Artists Rights Society, New York.

De Kooning began *Special Delivery* and *Secretary* with drawings in charcoal applied directly to unprimed sheets of coarse-textured wood pulp paper, portions of which are still visible around the edges of each picture and through some of the overlying paint. The artist then traced, reinforced, and expanded upon the preliminary charcoal designs, outlining the geometric shapes and anatomical forms with a fluid, black paint, probably applied with a liner's brush.[29] Using such a brush and possibly turning the paper as he worked, he created in *Special Delivery* the intricate network of overlapping shapes that appear to weave in and out and across the picture plane. In *Secretary*, drips in black paint near the top center that flow upward toward the top edge and the adjacent trickles of medium that flow downward through the yellow paint indicate that the artist also executed this work from different directions.[30]

Once de Kooning laid in the initial charcoal and black paint designs in *Special Delivery* and *Secretary*, he applied additional layers of paint, alternating them with more charcoal. While the paint layers were still soft, the artist continued to work them, pushing and pulling brushes and possibly a scraper through the paint, using vigorous heavy-handed strokes. As he worked, the solvents in the wet paints softened and partially dissolved the lower black paint layers. The mechanical action of the brushes and scrapers with which he applied the upper paint layers further induced the black to "work up" into the top paint layers, staining them to sooty-looking colors.

De Kooning's use of paints that dissolve and bleed into one another has a counterpart in his richly varied and unorthodox use of charcoal. In *Secretary*, the artist bore down heavily on the broad, soft charcoal sticks to outline forms and create a frottage-like effect. Elsewhere, he smeared the charcoal stick into the still-soft paint, churning the black fibers into the paint to produce passages that appear rough and scoured. In both works, he dragged a sharp, pointed charcoal stick directly through the wet paint, scoring the surface to expose the underlying layers. As in *Queen of Hearts*, he scumbled charcoal into the paint and partially erased drawing elements to suggest alternative positions for shapes and forms and to make it apparent that his working method involved repeated revisions to his compositions.

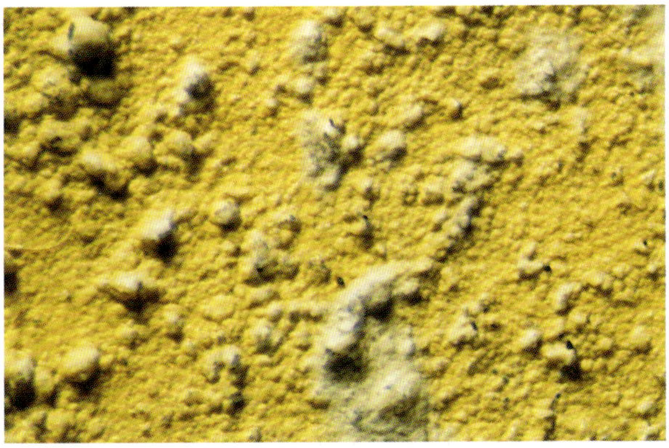

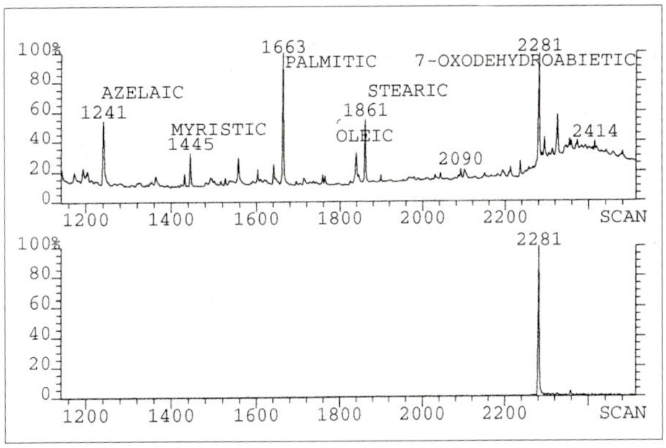

FIGURE 10
Photomicrograph of the lower right corner of *Special Delivery*, showing clumps of white material incorporated into the yellow paint (19x magnification).

FIGURE 11
Top: chromatogram (analysis by GC/MS, see Appendix 2) of a sample of yellow paint from *Special Delivery*. Early peaks in the series show the methyl esters of a drying oil. Later peaks show the oxidation products of abietic acids of pine resin (colophony). Bottom: scan for mass 328, the molecular ion of 7-oxodehydro-abietic acid.

Technical analysis of *Special Delivery* and *Secretary* reveals that the encrusted surfaces of these paintings, a key element in de Kooning's technique throughout the late 1940s and into the 1950s, are due as much to the types of paints that de Kooning employed as they are to his unusual methods of applying them. Examination of the surfaces of *Special Delivery* with the stereomicroscope shows, for example, that the rather gritty, granular surfaces of this painting result from irregular agglomerations of a white material that are partially embedded in both the yellow and white paints (fig. 10). This material has been identified as powdered plaster of Paris (calcium sulfate).[31] In the case of yellow paint, de Kooning mixed powdered plaster of Paris into retail trade paint. The chromatogram (fig. 11) of a sample of the yellow paint is typical of a drying oil with a methyl palmitate to methyl stearate peak area ratio (P/S) of 1.9, indicating it may be linseed oil and is certainly a drying oil (Mills 1966).[32] The later part of this chromatogram shows a series of large peaks that have been identified as oxygenated forms of the abietic acids of pine resin, often called rosin or colophony. The presence of these acids indicates that the medium of this yellow paint is a mixture of a drying oil with resin from a conifer, probably pine (Mills and White 1977, 14–16). Such a mixture, often termed an oleoresinous paint, was sold by the commercial paint industry as inexpensive enamel paint (Heaton 1947, 197–98, 339–40).[33] De Kooning's use of colored retail-trade paints in *Special Delivery* contradicts Thomas Hess's assertion that the artist restricted his use of house paints to black and white.[34]

The white paint in this work is evidently one that de Kooning prepared himself from plaster-of-Paris powder, but now mixed with an oil-resin varnish, rather than paint. This observation is supported by the fact that the plaster of Paris is the sole pigment present.[35] Additionally, the chromatogram of a sample of the white paint is similar to that of a naturally aged seven-year-old oil-resin varnish made from a mixture of polymerized linseed oil and colophony.[36]

Analytical evidence that de Kooning mixed plaster of Paris with his paints and used retail trade paints is supported by testimonial and archival evidence. A photograph taken in 1950 by Rudolph Burckhardt of de Kooning's studio shows a box with a Rainbow label sitting near his worktable amid cans of paint (fig. 12). Rainbow is the brand name of a line of dry products, including plaster of Paris, that was marketed by a New Jersey company beginning in 1946.[37] According to artist Patricia Passlof, who studied with de Kooning for several years beginning in 1948, de Kooning regularly mixed solid materials, such as plaster of Paris, into his paints to bulk them up and to extend his medium. Passlof also confirmed that de Kooning used a variety of retail trade paints, some of which were oil-based enamels; these he purchased from a nearby store that sold supplies for sign painting. According to Passlof, de Kooning knew these

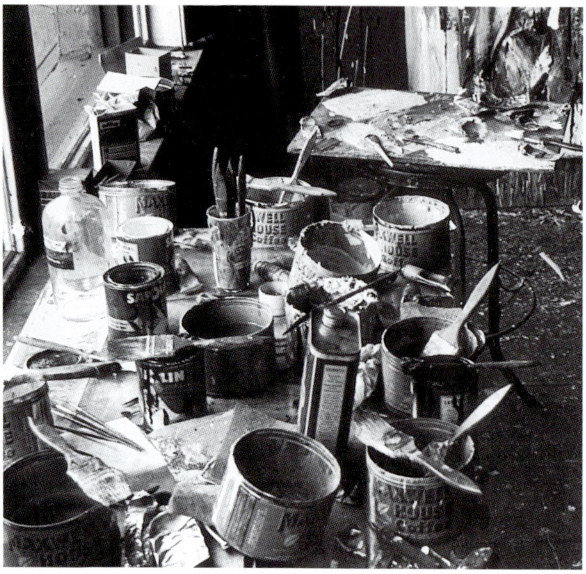

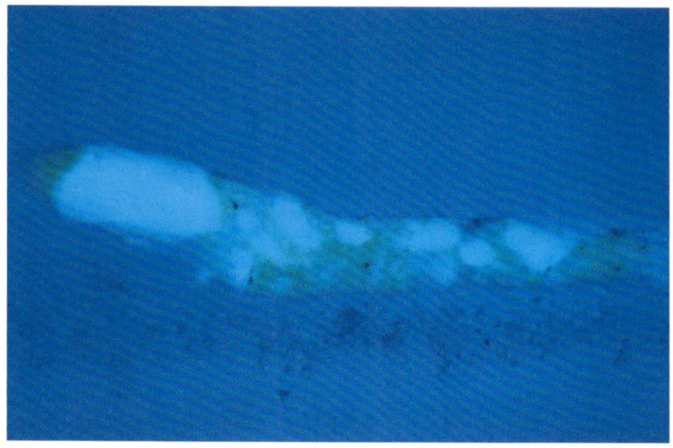

FIGURE 12
Interior of de Kooning's studio, New York, 1950, showing a box of Rainbow plaster of Paris on the windowsill and cans of paint with the Sapolin label on the worktable.

Photo © 2010 Estate of Rudy Burckhardt / Artists Rights Society (ARS), New York. Artwork © 2010 The Willem de Kooning Foundation / Artists Rights Society, New York.

FIGURE 13
Cross-section of white paint from the top right corner of *Special Delivery*, showing clumps of plaster of Paris powder unevenly distributed in the oil-resin medium. Photographed in ultraviolet light (250x magnification).

paints well from his days as a sign painter; they were far cheaper than artist's tube oils, and, the artist reasoned, should be extremely durable because they were designed for outdoor use (Passlof 1998).

While the low cost and ready availability of plaster of Paris, oleoresinous and enamel paints, and oil-resin varnish undoubtedly influenced de Kooning's practice of concocting his paints from these non-art materials, he also appears to have experimented with them in order to exploit their inherently gritty and porous qualities. The cross-sections taken from both the white (fig. 13) and yellow paints of *Special Delivery* elucidate the artist's motivation. These paints show that the white plaster material is massed across the picture surface in irregular clumps, while in any modern machine-ground paint—either artist and retail trade—the pigment particles would be homogeneously distributed in the binding medium. The inhomogeneity here is clearly due to a crudely prepared, studio-made paint.

It should not be supposed that de Kooning lacked the requisite training to make smooth, homogeneous paints. As an art student, he would have been taught how properly to prepare and mix paints by grinding the powdered pigment particles into the binding medium on a slab with a palette knife or muller until all the lumps disappeared. Only with such careful preparation could the artist be assured that the pigment particles would be uniformly coated or "wetted" with the oil and evenly distributed throughout the binding medium. When de Kooning used plaster of Paris, he elected not to grind the dry powder into his binding medium to produce an oil paint that would be smooth and buttery in its consistency. Instead, he probably wielded a brush or a stick to stir the powdered plaster into the paint or varnish, using containers such as the coffee cans so ubiquitous in studio photographs of the period (see fig. 12). Whether intended or fortuitous, de Kooning's method of mixing rather than grinding paints ensured that the dry pigment particles remained massed in the paint as isolated aggregates, even when brushed onto the paper support.

The artist's choice of unprimed, porous paper support also allowed him to accentuate the coarse texture of his paints. Without a ground or a heavy size layer to create a nonabsorbent surface, much of the excess oil and resin in the paint was absorbed into the porous support. De Kooning's choice of coarse supports during the late 1940s was only partly, if at all, dictated by economic necessity. As has been shown, during

FIGURE 14
Photomicrograph of the bottom edge center of *Secretary*, showing the textural extremes of the paints (25x magnification).

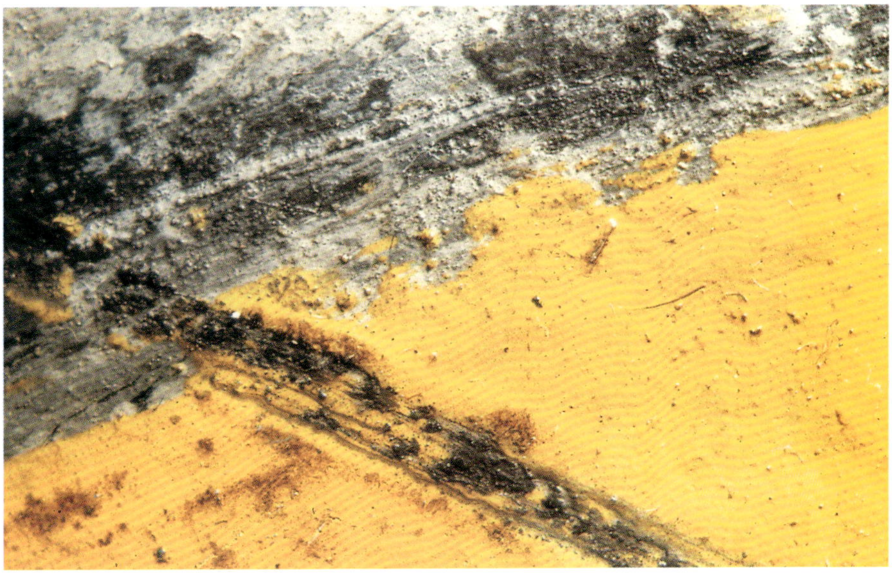

FIGURE 15
Secretary. Left: cross-section, gray paint, lower left corner, showing quartz particles and glass fragments incorporated into the paint film. Right: photomicrograph of the glass fragments found in the gray paint, lower left corner. Photographed under crossed-polar illumination (500x magnification).

the first half of the decade, when he was equally poor, he executed a number of paintings, including *Queen of Hearts*, with smooth grounds on the smooth side of Masonite panels. These unprimed, porous supports, while equally inexpensive, allowed the artist to expose the plaster-of-Paris particles in his paints as irregular clumps that protruded from the dried film, accentuating the course, gritty character of his inexpensive paints.

The textures of *Secretary* are still more extreme than those of *Special Delivery*, with slabs of coarse, gray paint acting as a counterpoint to the glossy black and the smoother yellows and orange (fig. 14). Pigment and medium analysis of this painting reveals that de Kooning again "doctored" his paints to enhance these textural contrasts. The black enamel is an inexpensive retail trade enamel that is a mixture of colophony with a drying oil. The binding medium of the smooth, almost greasy-looking yellow paint, however, is probably an artist's paint to which de Kooning added significant amounts of wax.[38] The chromatogram for a sample of this paint shows, in addition to the azelate, palmitate, and stearate peaks of a drying oil, a series of regularly spaced peaks that have been identified from their mass spectra as saturated C_{22} through C_{33} straight-chain alkanes. The configuration and relative intensities of these even- and odd-numbered alkanes correspond to reference samples of a paraffin wax, commonly used for making household candles and as a sealant for food products (White 1978, 57–68). Whatever its source, de Kooning surely incorporated wax into an artist's oil to produce paint with a slick, almost greasy, surface.

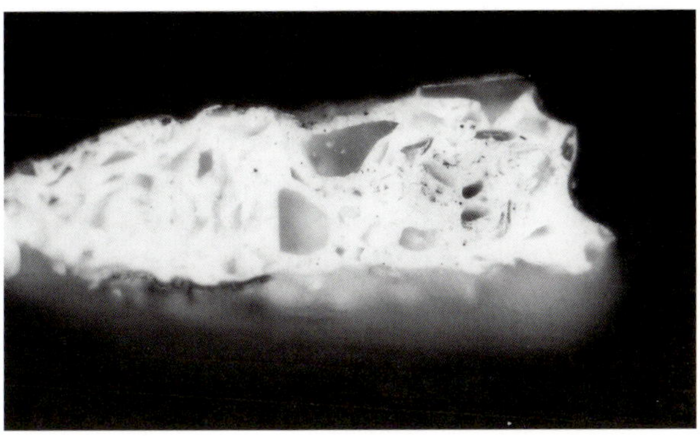

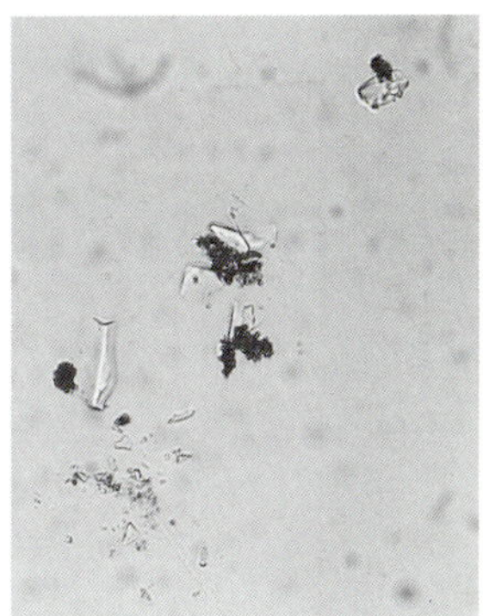

The adjacent gray paint, by contrast, is very matte and frankly scabrous in appearance. The paint itself has been identified as an oil-modified alkyd house paint, based on the presence of dimethyl phthalate as one of the major saponification products. Identification of the pigments and additives—titanium dioxide and carbon black in addition to what appears to be clay ($Al_2Si_2O_5(OH)_4$ and iron oxide)—the latter probably added by the manufacturer as a flattening agent—indicates that this paint was probably sold as a priming or flat wall paint.[39] The coarse texture of the gray paint is due also to the artist's addition of large amounts of irregularly shaped pieces of quartz, which can be seen poking up conspicuously through the surface of the paint film. De Kooning also added ground glass to the paint and these shards further intensify the coarse texture of the painting. Both the quartz particles and the glass fragments are clearly visible embedded in a cross-section and in dispersed paint samples of the paint film (fig. 15).[40] Inorganic analysis of two additional paintings by de Kooning from the late 1940s confirms that during this period the artist regularly added extraneous, gritty materials to his paints. Paint samples of *Valentine* (1947) revealed the presence of large particles of quartz. The coarse white paints of *Untitled* (1947–48) contain ground glass in addition to gypsum and calcite.[41]

This detailed examination of the paint surfaces of *Special Delivery* and *Secretary*—rich and smoothly fluid in some cases, scored and granular in others—and analysis of their complex paint structures leaves little doubt that de Kooning deliberately created the exaggerated textures of these paints, an effect described by Thomas Hess as "scarred" and "clotted" (T. Hess 1968, 45). Thus, the novelty of these two paintings and others from the same time period lies in their morphology as well as their imagery. In both *Special Delivery* and *Secretary*, de Kooning not only took his use of traditional art materials to extremes, he also experimented with a wide range of non-art materials and consciously exploited them to create a variety of visual impressions. As analysis has shown, by the last half of the 1940s the artist relied increasingly on the expressive qualities of his "doctored" paints to serve as the vehicle and bearer of his artistic vision. These paintings thus reflect de Kooning's intense response to his immediate urban surroundings, its buildings and streets, its light and grime. In several interviews, the artist, in fact, acknowledged that he tried to capture in his paintings the feeling evoked by his surroundings. "Even abstract forms must have a likeness," he stated (T. Hess 1968, 45).

Black-and-White Paintings: *Zurich* (1947)

Sometime in late 1946 or early 1947, de Kooning produced his first all black-and-white paintings, and it is these works that garnered him his first public recognition. In abstractions such as *Zurich* (fig. 16), painted in 1947, the artist forged an original style in which he exploited his painting materials, demonstrating a marvelous balance of technical improvisation and control.[42]

To a greater extent than any of the paintings discussed thus far, *Zurich* demonstrates de Kooning's improvisational working methods when applying his paints. The paint surfaces appear to be freely developed, without deliberate planning and with no exact order of working. Nevertheless, certain general and specific observations about de Kooning's working methods can be made, based on the study of *Zurich*. Over a preliminary drawing and most of the unprimed, wood-pulp paper support, the artist applied a brush coat of a glossy black paint so viscous that he probably used it directly from the can. As he did so, he allowed it to pool as a substantial layer on the surface of the paper,

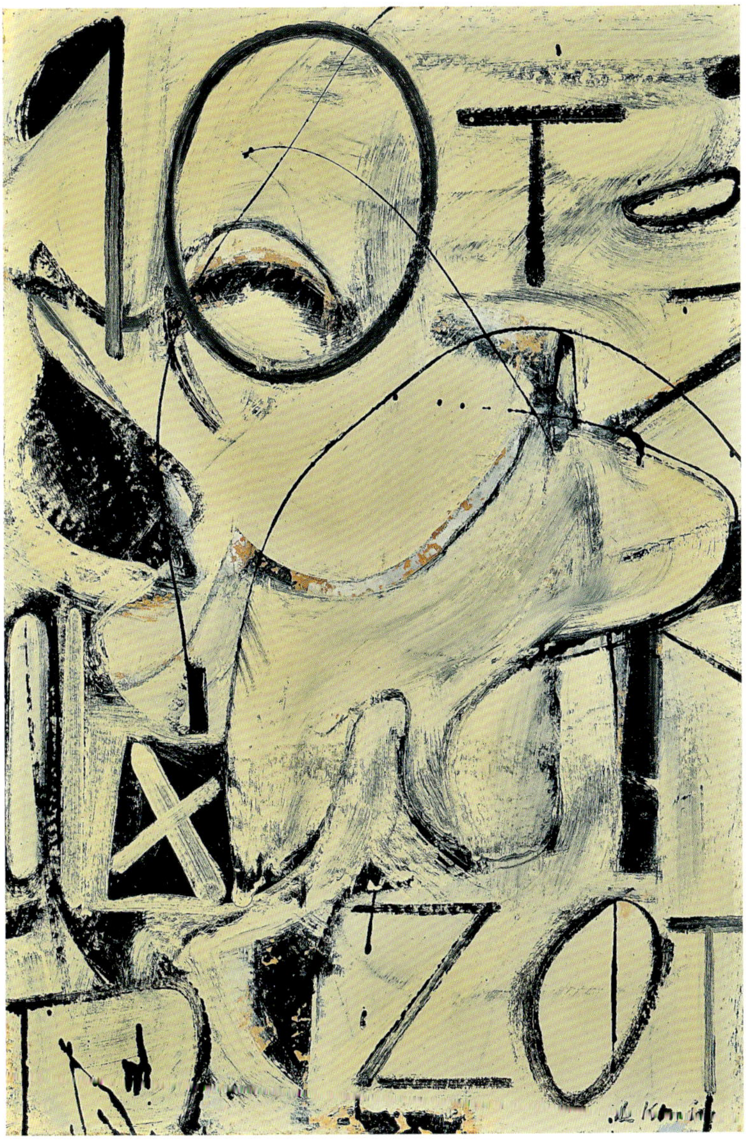

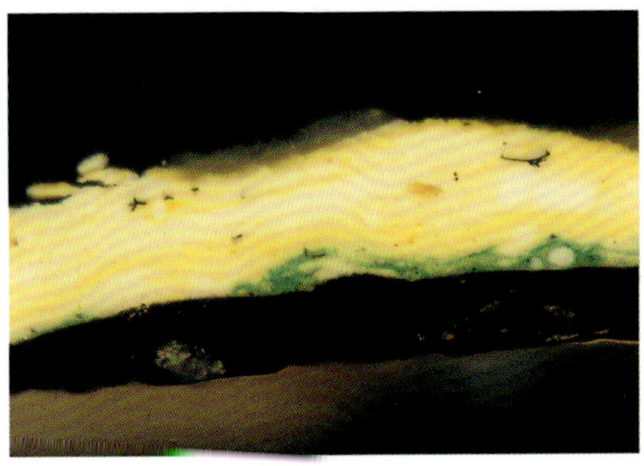

FIGURE 16
Zurich, 1947. Oil on paper, mounted on fiberboard, 91.4 × 61.3 cm (36 × 24⅛ in.). Hirshhorn Museum and Sculpture Garden, Smithsonian Institution (86.1355). The Joseph H. Hirshhorn Bequest, 1981.

© 2010 The Willem de Kooning Foundation / Artists Rights Society, New York. Photo: Lee Stalsworth

FIGURE 17
Paint cross-section from the center of the left edge of *Zurich*, showing three distinct paint layers (64x magnification).

where it dried to a smooth, high-gloss sheen. He then applied swirls of a creamy white paint over the enamel layer in rapid, fluid strokes. As he worked, moving different-sized brushes in all directions, he left voids or reserves in the white paint that reveal the lower black enamel as letters and abstract forms. He then applied additional black enamel, which he allowed to drip and smear as he drew it directly into the still-wet white paint, spelling out additional scattered letters and the word "ZOT."

The appearance of gestural immediacy and unpremeditated spontaneity in *Zurich* is, to some degree, an illusion. Despite the presence of a true wet-into-wet technique, evident in some of the upper paint layers, paint cross-sections show that the initial black and the subsequent white topcoat were built up as distinct layers; the black layer was undoubtedly dry before the upper white paint was applied (fig. 17). The impression of a wet-into-wet painting method actually derives from the artist's use of paints that could be readily redissolved. When de Kooning applied the white paint to the surface of the lower black enamel, the solvents used to dilute the white partially solubilized the lower black. And, as the artist applied the paint, the mechanical action of his brush worked the lower black up into the upper layer, which shows as dark streaks in the white. The two paints at first slightly repelled one another, but then they dried with a lava-like or marbleized effect, creating the deceptive impression—an impression the artist obviously appreciated—that the white and black paints were intermixed wet-into-wet on the painting support before either had set.

Significantly, the paint cross-section and pigment analysis of the white paints from *Zurich* and other works of this period, cited above, contradict John Brealey's assumption that the white paints in these paintings have yellowed with time. The distinctly warm tone of the white paint in *Zurich* is not due to discoloration of the medium, as has been suggested, but to the presence of small amounts of the pigment cadmium orange and a yellow organic dye that is readily visible swirled into the white.[43] Likewise, the warmer tones of the off-white paints of both *Valentine* and *Untitled* (1947–48) are due to the presence of small amounts of organic red and yellow pigments, as well as iron oxides and carbon black.[44]

The identity of the paints in de Kooning's black-and-white paintings has been the subject of some conjecture. Thomas Hess identified de Kooning's enamel paint, ambiguously called synthetic enamel, by the brand names Ripolin and Sapolin, although it is unclear whether he is referring to the black paint, the white paint, or both.[45] Several cans of paint with Sapolin labels are visible on the worktable in de Kooning's studio in the 1950 photograph by Rudolph Burckhardt (see fig. 12). Hess also wrote that de Kooning used the same paints in the black-and-white paintings as in his black-and-white drawings of the same period (T. Hess 1972c, 13). Conrad Fried, the artist's brother-in-law, acknowledges that the blacks in these works are retail trade paints; however, he maintains that the artist knew that white enamel house paint would discolor with time, so he used an oil-based medium with zinc white pigment for the white paints in these pictures (Prather 1994, 103, no. 15). Patricia Passlof (1998) recalls that the blacks that de Kooning used in these paintings were Ronan's drop black quick-dry lettering enamel and that the whites were paste white house-paints bought at Behlen Brothers paint store. Behlen Brothers sold both Ripolin and Sapolin paints.

Medium and pigment analysis of the white paint of *Zurich* revealed that it is an oil-based paint, although it is not an artist's paint as Fried's comment implies. Instead, it is an inexpensive oleoresinous paint made from a mixture of linseed oil (P/S 1.7) with a large amount of colophony. The high proportion of abietic acids from rosin in the sample is significant, indicating that the percentage of colophony in the original paint was greater than that of the oil. The paint, therefore, probably was marketed as enamel or high-gloss paint. The pigment composition of the paint is predominantly titanium dioxide, with only small amounts of zinc white.[46]

The binding medium of the black paint of *Zurich* was not conclusively identified but it is probably one of the several modified oleoresinous paints that were developed around mid-century. Pyrolysis-GC/MS analysis of samples of the glossy blacks from both *Zurich* (1947) and *Black Friday* (1948) revealed nearly identical pyrograms with saturated C_{14}, C_{16}, and C_{18} methyl esters, in addition to a series of large peaks that eluted later in the run. These peaks were identified as twenty-carbon ring structures with mass spectra base peaks identical to the various oxidized abietadiene acids of colophony, but with molecular ions from 16 to 45 mass units higher than those acids. Although further analysis is still needed, these enamels may be an oil-ester gum paint made from esterified colophony combined, in the presence of heat, with a drying oil. Sometimes called four-hour enamels, these high-gloss paints were much faster drying and more durable than the more common mixtures of untreated colophony and oil, becoming "touch dry" within three to four hours after application. Four-hour enamels

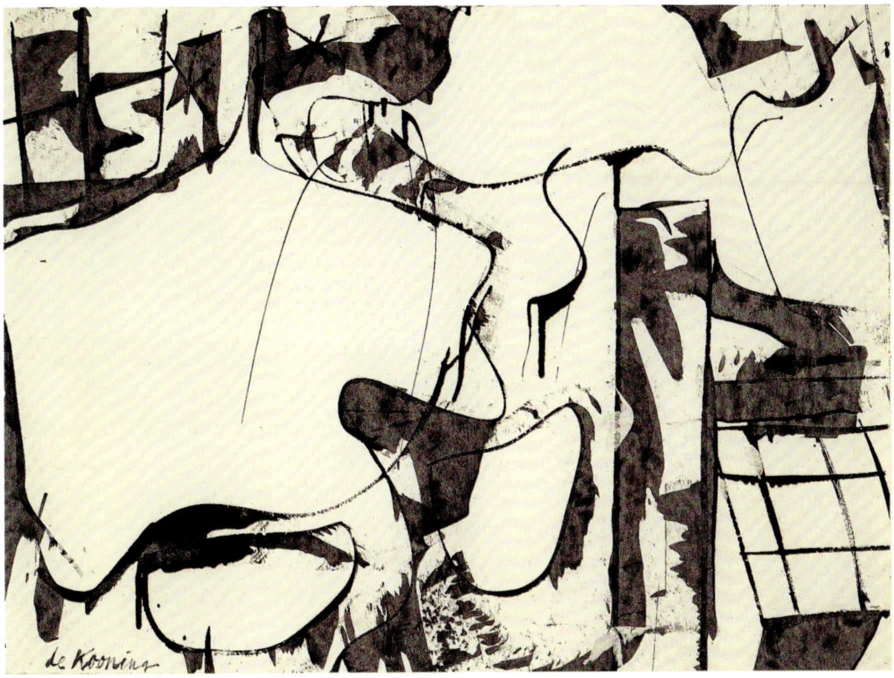

FIGURE 18
Untitled, 1949. Ink on paper, 55.7 × 76.2 cm (21 11⁄16 × 30 in.). Hirshhorn Museum and Sculpture Garden, Smithsonian Institution (66.1220). Gift of Joseph H. Hirshhorn, 1966.

© 2010 The Willem de Kooning Foundation / Artists Rights Society, New York. Photo: Lee Stalsworth

were designed primarily for spray application, although they could also be applied with a brush or roller.[47] The black paint from the drawing *Untitled* (1949; fig. 18) was identified by pyrolysis-GC/MS as a protein-based ink rather than a drying oil, contradicting Hess's assertion that the black enamels in the black-and-white paintings and the drawings are the same.

De Kooning in a 1959 interview said that he used retail trade paints in the black-and-white pictures because they were inexpensive: "I did it because I wanted to be free of the material. I did not have any money, but I could go to a store and buy a gallon of white and a gallon of black and be in business."[48] Impoverished as he was in the 1940s, he presumably could not buy even the cheapest artist's oil paint and, concerned that the colors in commercial paints would fade, he limited his palette to black and white. Nonetheless, before and after creating the black-and-white paintings—when he was equally poor—de Kooning produced abstractions in color and, as this study has shown, for colored paints he used both artists' oils and retail trade paints. It is likely, therefore, that lack of funds was not the only reason de Kooning temporarily abandoned the use of color and used retail trade paints. He must also have been seeking a way to explore the expressive potential of forms without the added distraction of color. He later confided to a friend that he had worked in black and white because, among other reasons, color is "a great seducer" (Finkelstein 1967, 29), confirming Barnett Newman's observation that black and white are the "tools for radical experiment" (T. Hess 1972c, 33). Many of de Kooning's colleagues—Barnett Newman, Jackson Pollock, Robert Motherwell, Arshile Gorky, Adolph Gottlieb, and Franz Kline—experimented during the late 1940s and early 1950s with paintings restricted to black and white. In his article on the black-and-white paintings of the Abstract Expressionists, Lawrence Alloway concluded that the connection of black with message-making was part of a widespread postwar effort to invest abstract art with a momentous subject (Alloway 1960, 49–62).

Just as de Kooning exploited the inherent properties of his painting materials to convey a sense of immediacy and a lack of premeditation that is not altogether

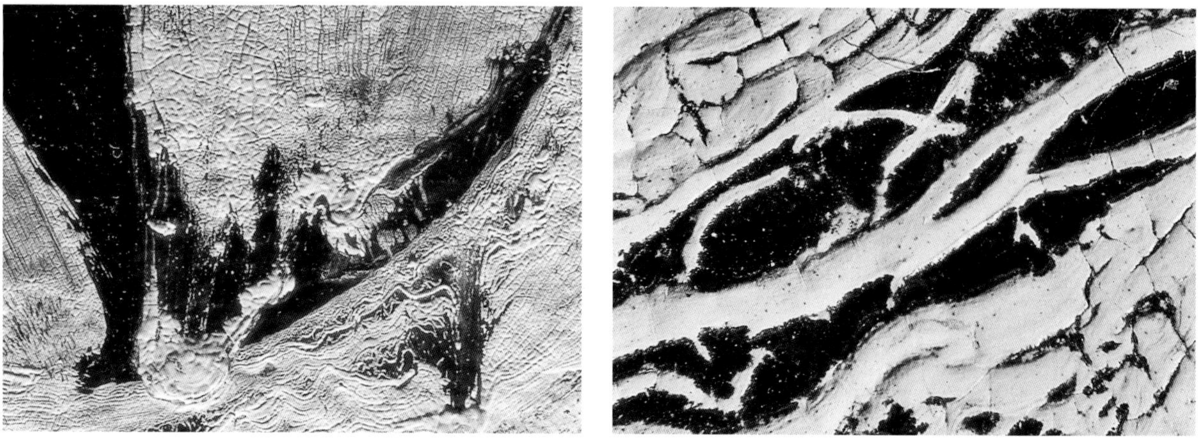

FIGURE 19
Zurich. Left: detail, below and left of center, showing the distinctive drying crackle pattern in the upper white paint layer. Photographed in raking light. Right: photomicrograph of the crackle pattern (13x magnification).

ingenuous, he deliberately "misused" his materials in a clear attempt simultaneously to harness accident and fight technical control. Examination of the surface of *Zurich* reveals that the upper white layer of oleoresinous paint has developed a fine "alligator" crackle pattern over most of its surface (fig. 19). This distinctive crackle pattern usually occurs when the upper coat of paint is less elastic or flexible than the underlying layer or when the lower paint layer has been prepared improperly. Art manuals caution artists that paints of high oil absorption, such as carbon blacks, should never be painted over with pigments of lower oil content such as zinc oxide or titanium dioxide, because drying cracks frequently occur within a short period of time in the relatively inelastic upper film. The addition of large amounts of colophony to the white paint makes the paint increasingly rigid and further enhances the probability of rapid drying crackle. Cracking is still more apt to occur when the upper paint layer is applied over a smooth, hard surface that does not provide a "tooth," that is, a properly roughened surface to which the upper layer can develop a secure adhesion (Mayer 1970, 172). De Kooning clearly ignored all of these basic painting principles when he created *Zurich*. In the absence of a strong bond to the smooth underlying layer of black enamel, the more brittle upper white contracted and cracked upon drying, forming molten, round-edged islands that sloughed down the face of the painting and merged with adjoining areas of soft paint as they ran.

It is impossible to determine with certainty whether the peculiar paint surfaces of *Zurich* were planned or constituted a happy accident. The visual effects—effects that probably manifested themselves shortly after the work was executed—are as likely to be the result of improvisation as they are of careful calculation. The characteristic crackle pattern in *Zurich* is, however, a phenomenon familiar to house painters. Known as "cissing," this cracking system commonly results when an oil film is applied over a lacquer or an enamel layer (M. Hess 1965, 98). Both cissing and alligatoring are common faux decorative finishes that de Kooning probably learned as a student and which he would have used while working for various decorating firms. Thus, it is probable that de Kooning, who proudly described himself as a good house painter and "not a smearer," deliberately departed from the time-tested rules of paint application as he simultaneously utilized many tricks of his trade, knowing that his transgressions were likely to produce intriguing and unforeseeable results (Shirey 1967, 80).

In *Zurich*, to a much greater degree than in previous pictures, de Kooning employed automatic techniques and took advantage of accidental effects. At the same time, he demonstrated an intimate familiarity with his materials. As he did in the colored abstractions, he selected from a variety of non-art materials, exploited their inherent qualities, extended their boundaries, and pushed them to produce unusual effects. As Paul Cummings aptly observed, de Kooning pushed his trained talents beyond their traditional limits, "courting disaster to discover a fresh statement" (Cummings 1983, 21).[49]

Even in a painting as abstract as *Zurich*, there is evidence of de Kooning's careful observation of the real world: fragments of lettering as well as shapes, colors, and textures are reminiscent of the proliferation of visual information randomly glimpsed on the evening streets. Aptly described as "urban nocturnes," these dense, compact compositions are filled with hints of reality, from peeling paint on asphalt to graffiti-covered walls and cracks in the sidewalks (Zilczer 1993c, 35).

Women Paintings: *Woman* (1948), *Woman* (1953), and *Two Women in the Country* (1954)

Woman (1948)

Woman (fig. 20), painted in 1948, is the first major surviving figure painting from the period in which de Kooning was primarily producing black-and-white abstractions. Composed of sweeping passages of paint applied to a panel of Masonite, the artist's aggressive distortion of the figure is unsurpassed among his earlier figurative works and foreshadows his women and abstractions of the 1950s. As with previous paintings, the effect of pictorial immediacy belies the complex methods that went into its production. While it appears to be a rapid, spontaneous rendering, close examination shows that de Kooning must have worked on it in several stages and that he selected and exploited his paints to underscore his imagery.

The female subject of *Woman* (1948) sits in a contraposto position wearing a short pink skirt. She flashes a toothy grin as she turns her head, bug-eyed with flowing yellow hair and flared nostrils, toward a star located above her right shoulder. Composed of ragged, swiftly applied strokes of paint that simultaneously dissolve and define her figure, the woman appears amid slashing, freely brushed passages that break through and penetrate the discontinuous contours of her form. The figure's three-dimensional presence within space has been forced to yield to the overall unity of the painted surface. While de Kooning's later statements implied that his paintings of women are an irreverent homage to the age-old tradition of the female in Western art, they are also rooted in urban vulgarity and popular culture: the grinning females from magazine ads and billboards and the sex goddesses of calendars and movies. Sidney Geist, describing *Woman I*, wrote, "Her image exists in the vast area between something scratched on the wall of a cave and something scratched on the wall of a urinal" (Geist 1953, 15). He might just as well have been speaking of *Woman* (1948).

Although much has been made of de Kooning's return to the figure in the early 1950s, an untitled black-and-white abstraction on the reverse of *Woman* (1948), and two unfinished figure studies on the back of the abstraction *Black Friday* (1948), provide concrete evidence that de Kooning was working simultaneously in both modes.[50] A detailed examination of *Woman* (1948) further supports the proposition that de Kooning's abstractions and figurative modes were not separate enterprises. He seems initially to have conceived *Woman* (1948) as a black-and-white abstraction, the anatomical elements having been added only late in the painting's execution. While x-ray examination of the painting was unhelpful, paint cross-sections taken from various locations throughout the painting all show a heavy buildup of paint in layers limited to a range of warm whites, grays, and a pure black (fig. 21). In each of the samples, the colored paints used to render the female figure and the off-whites around the star in the upper left corner were added only in the final stages of the painting's composition.

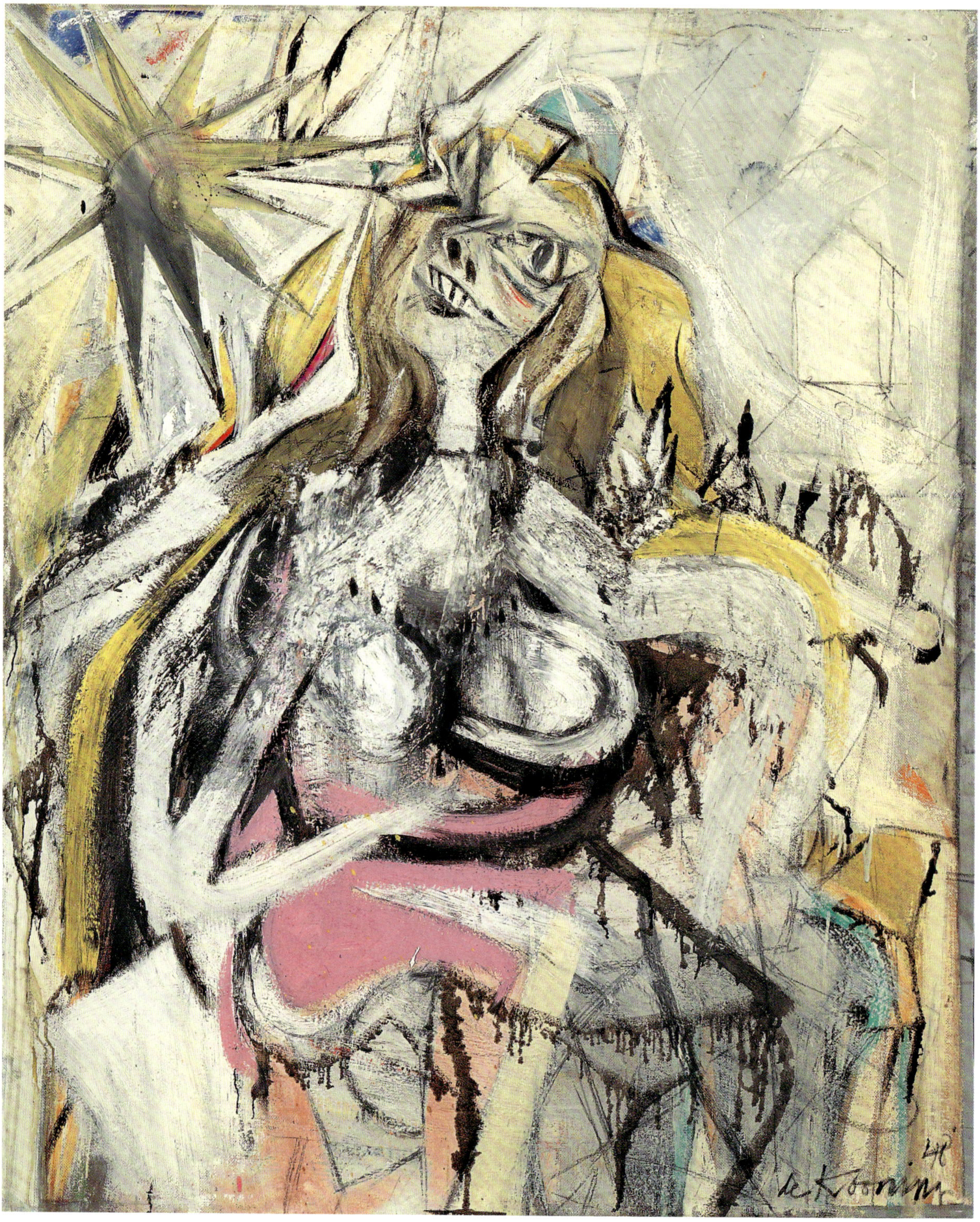

FIGURE 20 (opposite)
Woman, 1948. Oil on fiberboard, 136.1 × 113.2 cm (53⅝ × 44⅝ in). Hirshhorn Museum and Sculpture Garden, Smithsonian Institution (66.1197A). Gift of the Joseph H. Hirshhorn Foundation, 1966.

© 2010 The Willem de Kooning Foundation / Artists Rights Society, New York. Photo: Lee Stalsworth

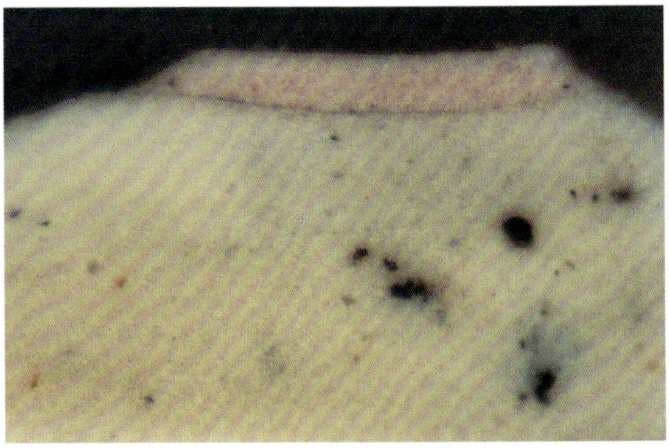

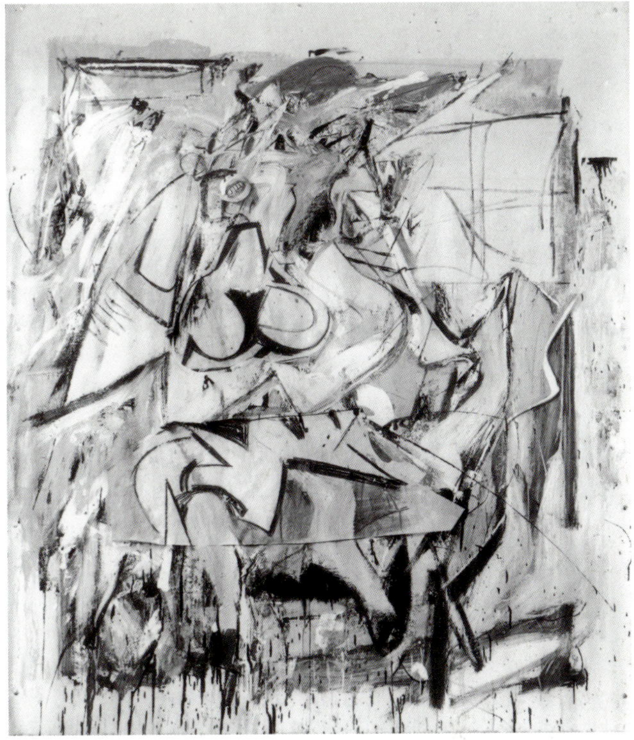

FIGURE 21
Paint cross-section from the pink skirt of *Woman* (1948), showing the lower paint layers of an earlier black-and-white abstraction with the pink of the final image added at the final stage (64x magnification).

FIGURE 22
State I of *Woman I*, 1950.

Photo © 2010 Estate of Rudy Burckhardt / Artists Rights Society (ARS), New York. Artwork © 2010 The Willem de Kooning Foundation / Artists Rights Society, New York.

Although no cartoons or studies survive for *Woman* (1948), a piece of masking tape still attached to the surface of the painting as well as straight edges and rippled impasto peaks in the paint surfaces indicate that de Kooning may have created the fragmented, disjointed figure by following a variation of the practice, described earlier, of placing drawings directly on the painting surface. De Kooning's use of drawings traced from other drawings and paintings means that elements from one work may reappear in other pictures in new combinations. *Woman* (1948), for example, is directly connected to the initial stage of *Woman I*, with her U-shaped breasts, jagged skirt, and the sweeping, disconnected contour of her left arm (fig. 22). These forms may derive from a single drawing or from a tracing made of *Woman* (1948) that was transferred to *Woman I* at an early stage in its development.[31]

Striations throughout the painting, but especially visible in the upper right corner, indicate that de Kooning was working in a well-established pattern of applying and then partially scraping down his paint. In *Woman* (1948), he scraped and possibly sanded much of the surface of the initial black-and-white painting prior to beginning the current image of the figure, although a significant build-up of paint layers from the original composition remains. Thus, the original black-and-white composition in muted tones of white, black, and gray forms the base for the image of the woman. The recessed area of the star in the upper left corner, for example, is a portion of an initial black-and-white layer that remains visible and that serves as part of the design of the final composition.

Examination of *Woman* (1948) also indicates that, although parts of the painting were completed with real speed and immediacy, the whole was conceived with calculation and control over an extended period of time. At first glance, the painting seems to have been done entirely in one sitting, with colors actually blended on the picture surface, wet paint into wet paint. And most of what we see was indeed painted at great speed with fluid paint and loaded brushes. The paint cross-section previously mentioned and examination of the surface with a stereo-microscope, however, provides evidence that the main areas of lower paint layer were dry before the uppermost paint layers were applied. A clear example is the woman's bright pink skirt, which was undoubtedly dry before the bold strokes in black and white were applied to its surface. This sequential approach to paint application seems at odds with the artist's graphically gestural style, but it was the logical consequence of a working method that included the use of fast-drying retail trade paints and a practice of painting in distinct stages with significant intervals of time between phases.

Many visitors to the artist's studio reported de Kooning's working method as being protracted rather than slow in speed. He worked in spurts of feverishly concentrated activity, separated by long periods of studying the image. His willingness to modify his painting in considered stages illustrates an element of deliberation that seems to belie the spontaneity implied by the words "action painting." Yet even Harold Rosenberg, who coined the term, acknowledged that de Kooning never used improvisational techniques exclusively: "Though they may start with a scribble or a sign, his forms are animated by conscious intuition. . . . Without being premeditated, his movements occur under the scrutiny of his aesthetic conscience" (Rosenberg 1963, 134).

A comparison of the pigments and the paint media of *Woman* (1948) with those of *Zurich* also suggests that there is less of a dichotomy between de Kooning's black-and-white abstractions and his figurative pictures of the period than previously thought (see Appendix 1, Table 1). The off-whites of *Woman* (1948) were executed with paints that have a pigment composition similar to those in *Zurich*: titanium dioxide, predominantly, with smaller percentages of zinc white and additions of cadmium yellow and orange, carbon black, and iron oxide, to give the white paint its warm tone. The binding medium of the white and off-white paints of *Woman* is also the same as that of *Zurich*: an oleoresinous paint that is a mixture of linseed oil (P/S 1.9 and 1.8) and colophony.[52] Significantly, the black paint in *Woman* (1948) is not an oleoresinous or an oil-ester gum paint, but an oil-modified alkyd, and this appears to have had important implications for the artist's working methods, as will be shown.

Like many works from the later 1940s, *Woman* is remarkable for its emphatic surface texture. The painting was executed on the rough side of a fiberboard panel with no ground or priming layer to diminish or conceal the prominent screen pattern of the particle board. The heavily textured surface of the support is clearly visible in much of the painting, where half-dried paint was dragged across the screened surface (fig. 23). As in other works of the period, de Kooning introduced coarse, granular materials into his paint medium to accentuate the rough texture of the support and the crude nature of his inexpensive paints. Paint cross-sections and microscopic examination of the paint surface show that the artist blended quartz, probably sand or small pebbles, into the gray paint near the lower right corner, and that he incorporated ground glass, brown in color and possibly from a beer bottle, into the white and pink paints around the seated figure, as can be seen in a photomicrograph and a cross-section (fig. 24). He also mixed significant amounts of plaster of Paris into the white paint used to render the torso of the figure and the star in the top left corner.[53] The addition of this powder contributed substantially to the coarse character of the paint. As in previous paintings, de Kooning rubbed charcoal onto the paint surface, deliberately smearing the black fibers into the paint to smudge and dirty the brighter colors. The textural augmentations from glass, quartz, plaster of Paris, and charcoal are compounded by the artist's addition of generous amounts of palette scrapings to many of the final paint layers. These large bits of dried paint, evident with the naked eye as a rough encrustation projecting through the paint, contribute significantly to the rugged, uneven appearance of the paint film.

FIGURE 23 (right)
Detail near the lower right corner of *Woman* (1948), showing the screened pattern of the fiberboard support, visible through the overlying paint layers.

FIGURE 24 (below)
Woman (1948). Top: photomicrograph showing shards of brown glass in the pale pink paint (32x magnification). Bottom: cross-section, near lower right corner, showing brown glass embedded in the white paint (250x magnification).

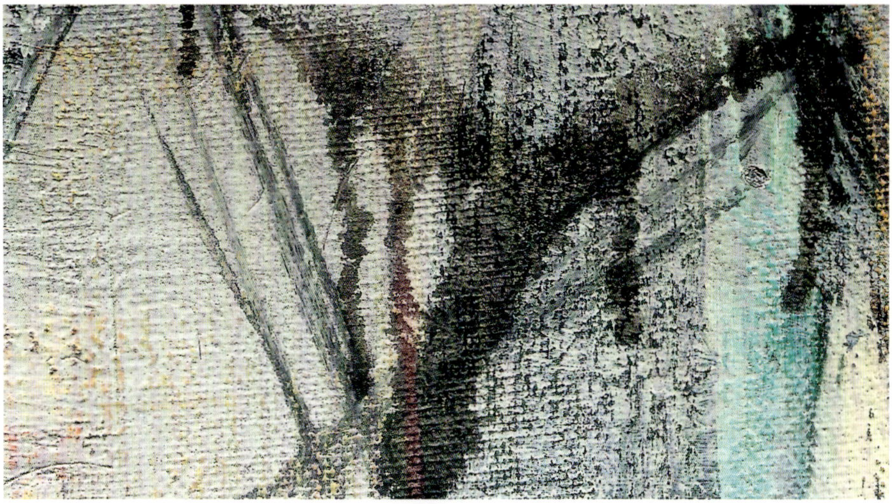

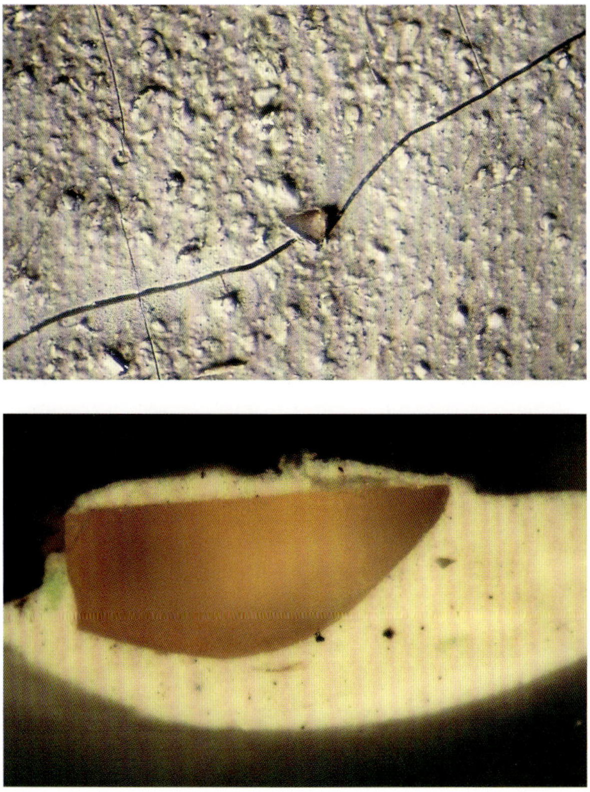

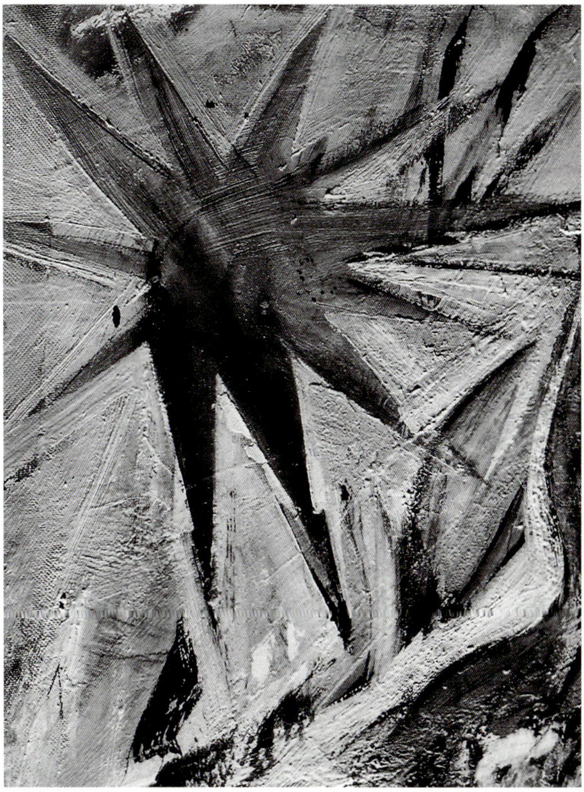

FIGURE 25 (above right)
Detail from top left corner of *Woman* (1948), showing the artist's build-up of paint, applied around the perimeter of the star with a combination of brushes and scrapers (photographed in raking light).

The pronounced texture of this work is due as much to the artist's methods of applying his paint as to the materials that he selected. In this work, de Kooning manipulated paints of vastly different consistencies, applying them with brushes of various sizes, palette knives, and scrapers to produce paint passages that vary enormously from one area to the next. This technique is a striking model of painterly freedom. Throughout the picture, paint is used to suggest substance as well as form and, in many parts of the work, the material nature of the paint itself overwhelms the details of the subject depicted. The artist, for example, schematically rendered the woman's breasts, shoulders, and multiple arms as well as the star in the upper left corner in opaque white house paint, made bulkier with the additional of plaster of Paris (fig. 25). The heavily applied brushstrokes and scrape marks throughout these areas are combined layer upon layer in almost sculptural masses.

The fluid black paint, used to bind the figure's breasts and torso, appears particularly to have been applied with little care. Significantly, de Kooning did not employ this paint as he had in the three earlier paintings discussed. That is, he did not apply it as a lower layer that he then softened with solvents and worked up into the overlying paints. Rather, this oil-modified alkyd house paint was one of the last paints applied to *Woman* (1948). The artist first diluted it to the consistency of a watercolor and then let it flow down the picture surface in runnels so that it stained the lower paints to a gritty black. Before it dried, he went back over the paint with a dry brush or knife, scraping and dragging his tools through the nearly dry lower white until the two colors, coarsely blended, assumed a scuffed, troweled appearance.

Identification of the paints in *Woman* (1948), as well as those used in *Zurich*, *Secretary*, and *Special Delivery*, again contradicts the assumption that de Kooning's paints, with few exceptions prior to about 1960, were conventional artists' paints, simply applied in unconventional ways (T. Hess 1953, 65). Rather, analysis reveals the extent to which de Kooning was taking advantage of the broad range of retail trade paints available during the 1940s: oleoresinous paints and enamels, oil-ester gums, and oil-modified alkyds. Analysis also reveals that de Kooning did not use any of the new synthetic polymer paints and their dispersions—the "latex" paints—that were becoming increasingly available around mid-century. Rather, he restricted his experimentation to a variety of different commercial paints that he purchased at different times and from different vendors, but which were all marketed as "oil paint."

By the late 1940s, most oil-based retail trade paints and enamels available to the public were, in fact, modified or synthetically prepared products and, as such, exhibited significantly different working properties.[54] Both oleoresinous paints that include a high percentage of untreated colophony and oil-ester gums dry initially due to rapid gelation of the polymerized oils and the evaporation of the solvents from the resin components. They remain freely soluble in practically all of the organic solvents for some time, usually several days; prior to that time, they can be redissolved readily and worked up with the application of successive coats of paint. Oil-modified alkyd paints, by contrast, cross-link within hours to form an insoluble paint film.[55] In the 1930s, linseed oil and other drying oils mixed with resins were still the most common medium for both interior and exterior architectural paints and enamels, but oil-modified alkyds were beginning to replace them. By 1940, alkyds were superseding oil paints, and in the early 1950s they had become the preeminent binding medium for architectural coatings.[56]

The oil-modified alkyd paints identified in *Woman* and *Secretary*, both dated 1948, mirror the changing paint industry. At the time when he was using oleoresinous and oil-ester gum paints and enamels, de Kooning's practice of dissolving and working up one paint layer into successive coats was a dominant feature of his paintings. Once the commercial paint industry began making the transition to alkyd paints, however, de Kooning's old method of redissolving his dried paint films was no longer possible. While he may not have been aware that he was using the less soluble alkyd paints in his pictures, de Kooning did refer to a new dilemma when he told authors Harriet Janis and Rudi Blesh that he was having difficulty with oil paints that do not "work up" as lacquer does (Janis and Blesh 1960, 31).[57] As medium analysis of the paints in the pictures of this period has shown, what de Kooning referred to as "oil" paint was seldom artists' paint; rather it was oil-based house and sign paint. By the late 1940s, these "oils" were more likely to be the more durable oil-modified alkyds rather than the oleoresinous paints. In *Woman* (1948), de Kooning's solution to the "dilemma" of the less soluble black enamels was to reverse his working procedure by applying the highly diluted black enamel as the upper, rather than lower, layer, and then mechanically working it into the lower layers with scrapers and brushes. As will be shown, de Kooning would soon develop alternative methods for slowing down the drying rate of his paints.

Woman (1948) has the quality of a bold experiment. Technical examination and analysis leave little doubt that de Kooning was embarking on a significant initiative of freer paint handling with this work. The final painting retains an appearance of immediacy, but the artist worked long and hard to achieve it. While the rough, unprimed support, the intensely contrasting palette, and the use of coarse-textured retail trade paints relate *Woman* (1948) to the colored abstractions and the black-and-white paintings of the period, the paint's quality of directness and immediacy may be seen as an exploration of the painting methods de Kooning was to develop in the 1950s and 1960s.

Woman (1953) and *Two Women in the Country* (1954)

By 1950, de Kooning was fully engaged in a succession of drawings and paintings of women. His renewed focus on the subject culminated in his third one-man exhibition, "Paintings on the Theme of the Woman," which opened at the Sidney Janis Gallery in March 1953. The show included six major paintings and sixteen smaller works on the subject. These smaller works included pastels, drawings in graphite and charcoal, and a number of small oils on paper. The artist's concentration on the figure continued into the middle of the decade, and between 1953 and 1955 the women also appeared in duos and occasionally in trios.

Woman (1953; fig. 26), which was executed on a sheet of rag paper, depicts a female clad in a bright yellow sundress with flattened torso, truncated legs, and schematically rendered arms and hands. Her wide staring eyes and mouth are comically duplicated below as a second set of large cross-eyed nipples and a toothy, red-lipped pubis. *Two Women in the Country* (1954; fig. 27) is painted on artist-primed linen. The painting shows two voluptuous, monolithic figures that fill most of the picture space. In both works, the frontally posed, full-length figures are simultaneously descriptive and abstract. A swath of unpainted, primed canvas along the bottom of *Two Women in the Country* and exposed unpainted paper along the right edge of *Woman* make the figures appear to float above the foreground.

Examination of these two paintings reveals that drawing is still integral to the development of his paintings. Microscopic examination and infrared photography of *Two Women in the Country* and *Woman* (1953) indicate that de Kooning began both of these paintings with broad armature-like sketches in charcoal over which he liberally applied intersecting, arcing strokes of paint that both transgress and reinforce the outlines of the initial drawings. He then applied more charcoal, glossy black paint, and additional strokes of colored paint that delineate women's faces, torsos, arms, and legs and serve to bring the images back from abstraction.

Gritty stains in the foreground of *Two Women in the Country*, erasures on the paper of *Woman* (1953), and scraped paint passages in both works indicate that de Kooning was continuing his practice of liberally applying layers of paint and then repeatedly scraping them down. As in other works, the apparent speed of de Kooning's brushstrokes and the haste of the paint application in the work of the 1950s are increasingly at odds with the protracted time it took to complete the works. Critic Henry McBride's 1953 review of de Kooning's paintings on the theme of "The Women" at the Sidney Janis Gallery would also aptly describe *Woman* (1953) and *Two Women in the Country* (1954):

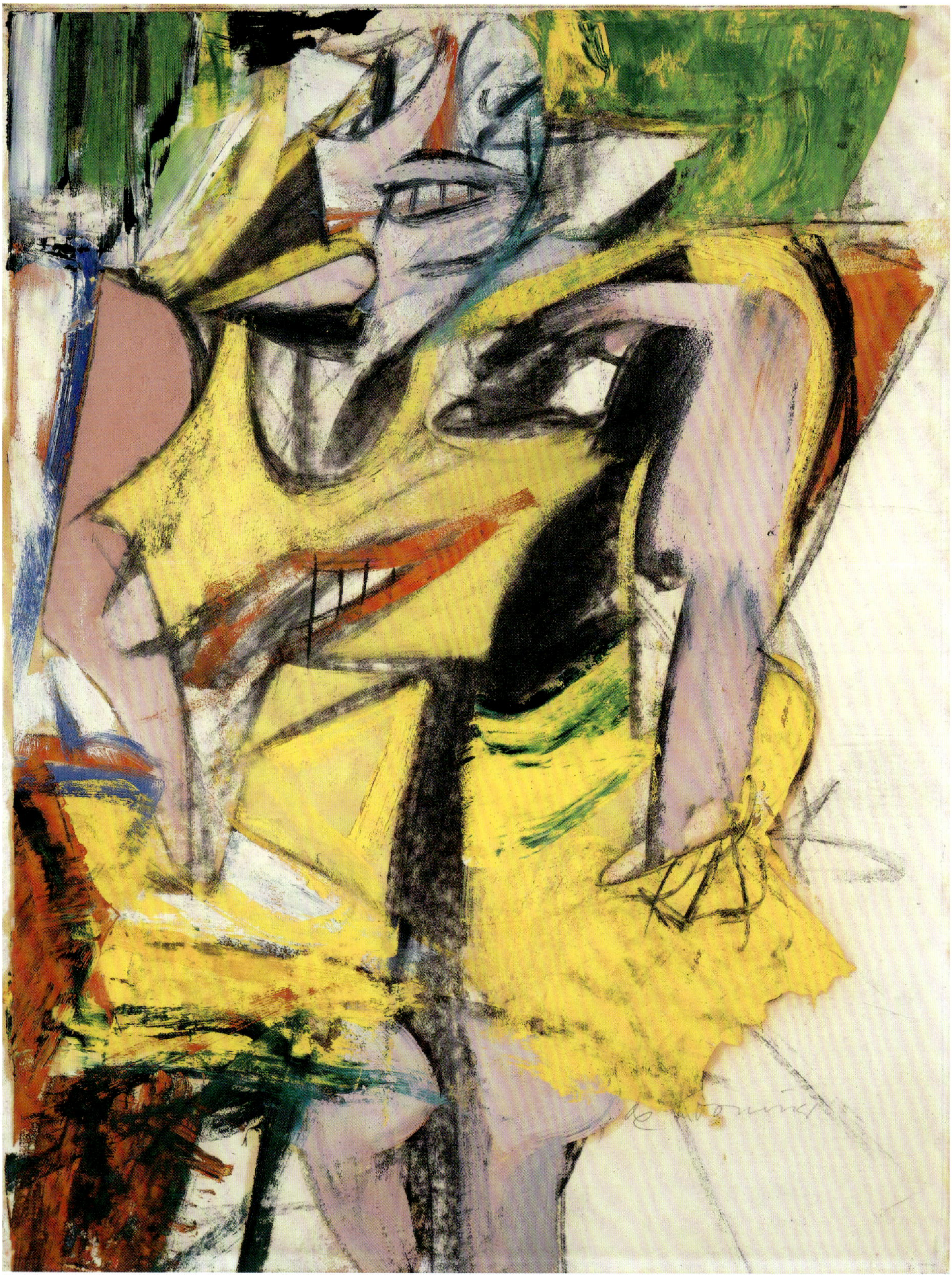

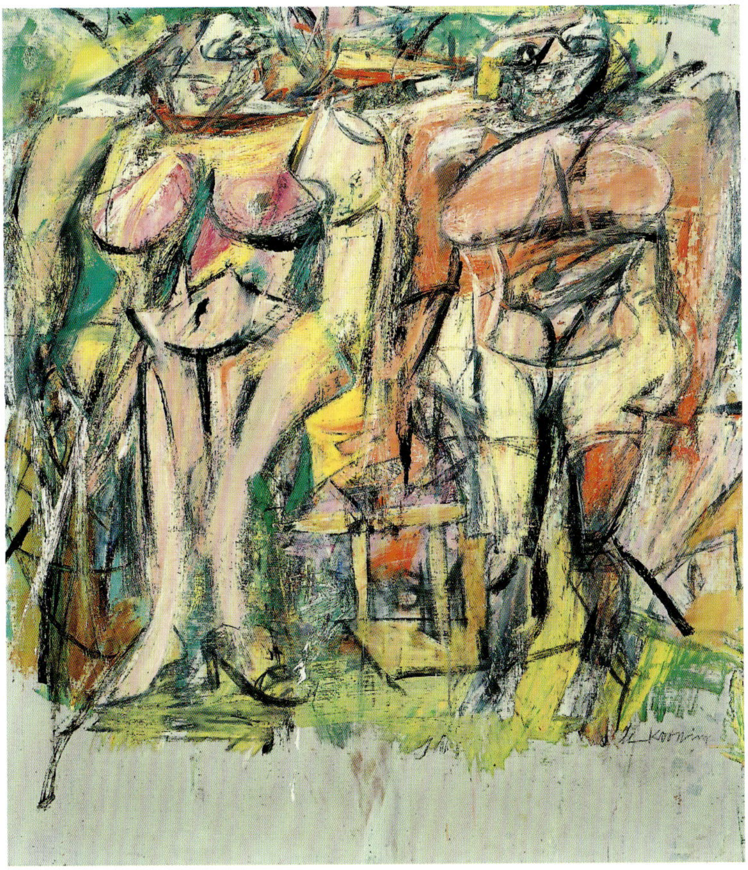

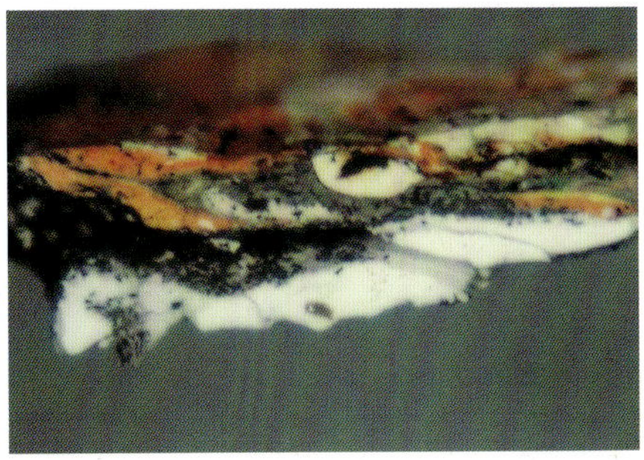

FIGURE 26 (opposite)
Woman, 1953. Oil and charcoal on paper, mounted on canvas, 64.9 × 49.8 cm (25⅝ × 19⅝ in.). Hirshhorn Museum and Sculpture Garden, Smithsonian Institution (66.1199). Gift of Joseph H. Hirshhorn, 1966.

© 2010 The Willem de Kooning Foundation / Artists Rights Society, New York. Photo: Lee Stalsworth

FIGURE 27 (above left)
Two Women in the Country, 1954. Oil and charcoal on canvas, 117.1 × 103.5 cm (46⅛ × 40¾ in.). Hirshhorn Museum and Sculpture Garden, Smithsonian Institution (66.1200). Gift of Joseph H. Hirshhorn, 1966.

© 2010 The Willem de Kooning Foundation / Artists Rights Society, New York. Photo: Lee Stalsworth

FIGURE 28 (above right)
Paint cross-section from the left edge of *Two Women in the Country*, showing the artist's incorporation of charcoal into his paints and his wet-into-wet working method.

these paintings, he wrote, "have not been dashed off hurriedly as one might think. On the contrary, the artist had struggled with them intensely… making endless studies and changes" (McBride 1953, 47).

Paint cross-sections taken from *Woman* (1953) and *Two Women in the County* indicate that de Kooning was changing his working method. He no longer altered his imagery by building up his paints in distinct stages, wet-over-dry, as he had done with both *Queen of Hearts* and *Woman* (1948). Nor was he relying on the solvent rich upper paint layers to redissolve the already-dried lower layers, as with *Secretary* and *Zurich*. Rather, the artist was now increasingly applying his paints wet-into-wet. A cross-section from the left edge of *Two Women in the Country*, for example, clearly shows the brushstrokes of one color mixed with the wet paint of another (fig. 28). Although distinct stages can be identified in both paintings, much of each work was carried out with great speed and fluidity, with brushstrokes that are actually superimposed, blurred together, and blended on the picture surface. Paint cross-sections from *Woman* (1953) and *Two Women in the Country* also demonstrate the extent to which the artist was fusing the practice of drawing and painting. In the cross-section discussed above, charcoal is embedded into the paint surfaces and thoroughly incorporated into each of the lower paint layers. As de Kooning worked on *Two Women in the Country*, his brushstrokes churned the charcoal fibers from the preliminary drawing into the overlying paint. He also dragged charcoal sticks directly through the still-wet paint surfaces and then added more paint over the upper drawing elements, further blending the charcoal and paint. As a result, the black, fibrous particles are thoroughly mixed with the paints and have become as integral to the paint structure as the colored pigments.

These samples provide physical evidence to support contemporary reports that de Kooning's pictures of the 1950s were increasingly referring to the act of painting itself. The artist used his paint extravagantly, brushing and dripping it on, layer upon layer. He would often pull spatulas through all or portions of the still-wet painting, scraping it down to the support, only to start over again. The residues of the rough, troweled surfaces become as much a positive gesture as are the areas of high impasto. As has been noted frequently, de Kooning's adaptation of a more expressive brushwork in the paintings of this period coincided with a major Chaim Soutine retrospective at the Museum of Modern Art in 1950.[58] De Kooning later acknowledged the extent to which he admired Soutine's visceral use of his materials and his identification of the physicality of women's bodies with the materiality of the paint: "I've always been crazy about Soutine.... Maybe it's the lushness of the paint. He builds up a surface that looks like a material, like a substance. There's a kind of transfiguration, a certain fleshiness, in his work" (Staats and Matthiessen 1977, 70).

Medium analysis and anecdotal evidence suggest that de Kooning, by the early 1950s, had begun to use artist's tube paints more frequently and only occasionally resorted to house or sign paints. The glossy black paint from *Two Women in the Country* is an oil-modified alkyd. As was the case in *Woman* (1948), this paint was one of the last to be applied and, in this case, the artist used it largely to define and outline the women's features. Samples of the white and orange paints from this picture, the orange, yellow, and green paints from *Woman* (1953), and two samples each from *Woman VI* (1953) and *Gotham News* (1955), reveal that the binding medium of these paintings from the 1950s is a drying oil, with none of the colophony, pentaerythritol, phthalate, or cellulosic components characteristic of retail trade paints of this period (see Appendix 1, Table 1).[59] With the exception of the white and orange paint samples from *Two Women in the Country*, the P/S area ratios of the paint samples are low. This may be explained by the presence of significant amounts of castor oil (a nondrying oil), along with a traditional drying oil, in the binding medium. This analysis supports narrative accounts that de Kooning had turned to artists' oils in the 1950s. By 1951, art dealer Sidney Janis had begun to advance the artist money to buy painting supplies. According to John McMahon, de Kooning's studio assistant at a later period, the paints that the artist was using by this time, and continued to use well into the 1970s, were Bellini artists' oils prepared by Leonard Bocour (McMahon 1993).[60] Undated formulations for Bellini oil paints show that the binding medium for these tube paints was a mixture of oils that included significant amounts of castor oil.[61] The presence of castor oil would have significantly slowed the drying rate of the medium, allowing de Kooning to rework his paintings for longer periods of time. It is interesting to speculate that de Kooning moved from retail trade paints and enamels to artists' oils in the early 1950s not solely because he could afford better paints, but also because he was increasingly dissatisfied with the faster drying alkyd house paints and preferred Bocour's formulations for slow-drying artists' paints. By the early 1960s, as will be shown, de Kooning would adopt alternative means of keeping his paints fluid and workable for longer periods of time.

"Painting Situations"

Contemporary critics and art historians have acknowledged a latent subject in de Kooning's paintings of the last half of the 1940s and the 1950s and a few have perceptively linked these paintings to the urban scene. De Kooning's friend Louis Finkelstein, an artist and writer, noted that "instead of painting objects, [de Kooning] paints situations"

(Finkelstein 1950, 205). Lawrence Alloway noted the "implicitly urban" nature of de Kooning's paintings of the late 1940s through the 1950s that "compress 10th Street and art history" (Alloway 1965, 20). Thomas Hess wrote that "the concept of the American banal has always intrigued de Kooning: advertising, commercial art. . . . He found it romantic, vulgar, and in its own way poignant . . . and he decided he liked this intrusion of the art of the streets into his work" (T. Hess 1968, 76). Irving Sandler observed that in their "restlessness claustrophobia, density, rawness, violence and ambiguity, de Kooning's paintings of the 1940s and 1950s feel like a walk down a Manhattan street" (Sandler 1993, 83). More recently, Judith Zilczer argued that de Kooning's paintings of these years stem from his enthusiasm for the visual chaos of New York City, "from torn billboards and discarded newspapers to the cracks in the sidewalk" (Zilczer 1993c, 33).

This in-depth study of seven paintings executed between 1946 and 1954 and the analysis of samples from other pictures of the period reveal the degree to which de Kooning adapted his painting methods and materials to his urban subject matter. Rough-textured supports are a prominent feature of his painting of this period, and the artist clearly felt that a certain coarseness of materials and technique was appropriate for an urban-inspired subject. His pictures include materials intimately associated with the urban environment—plaster, charcoal, ground glass, quartz, calcite, paper, fiberboard, and inexpensive enamels, sign paints, and house paints—as well as practices that incorporate scraping down, scoring, and smearing his materials to produce paintings that can be interpreted as visual metaphors for the gritty textures of Greenwich Village and the Lower East Side.

De Kooning himself repeatedly articulated his interest in the common, occasionally tawdry details of American life, and affirmed that the city streets were a major resource. In one interview he noted that, unlike many American artists who looked to Europe for culture, he deeply appreciated the dense squalor and the haphazardly encountered debris of the urban setting: "I was interested in signs, in letterings, things I noticed all around me in the city" (Parsons 1967–68, 4). Both Elaine de Kooning and their friend Edwin Denby recalled that the abstract shapes, colors, and textures of many of de Kooning's paintings derived from his observations of the messy vitality of the city and his fondness for the overlooked minutiae of the streets, expressed during their long walks through the neighborhood late at night, "The sidewalk cracks, gum-spots, the water, the bits of refuse" (T. Hess 1959b, 25).[62]

Yet visual clues to the local environment in de Kooning's paintings are purposely ambiguous. The imagery, colors, and texture of the paint may suggest a broad range of sensations, but in the end they are equivocal and unspecific. These paintings are informed by the artist's awareness that content or meaning had to be expressed through the paint itself rather than simply illustrated. For the symposium held at the Museum of Modern Art in 1951, the artist reflected, "Everything that passes me I can see only a little of, but I am always looking. And I see an awful lot sometimes" (de Kooning 1951b, 4). Indeed, the paintings of the late 1940s and 1950s convey the artist's keen observations of the urban milieu, viewed not from a distance but close up, evoking a personal response. Years later, when de Kooning and John McMahon attended an exhibition that included several of de Kooning's paintings of the late 1940s and early 1950s, McMahon commented on their dirty surfaces and asked de Kooning if they should be sent to a conservator. De Kooning simply replied that they did not need to be cleaned: they were painted to look that way (McMahon 1993).

3 Paintings of the 1960s

The art that Willem de Kooning produced after about 1960 is sometimes considered a distinct body of work.[63] These paintings clearly manifest a different style than those produced in the late 1940s and 1950s. As the New York paintings are infused with the feeling of the city, so the East Hampton works convey a sense of water and light and embrace the notion of the pastoral.[64] Ten years after his move to Long Island, de Kooning acknowledged that working in the country had affected his art: "I wanted to get back to a feeling of light in painting. . . . I wanted to get in touch with nature. Not painting scenes from nature, but to get a feeling of that light that was very appealing to me, here particularly, I was always very much interested in water" (Wolfe 1981, 16).[65] Technical examination and analysis of paintings of the 1960s reveals that in these works de Kooning continued many of the practices he used in paintings of the 1940s and 1950s, as he simultaneously developed new techniques and adopted new materials as a means of conveying his new environment. Thus, the works of this decade do not mark a sharp break with his earlier practices but rather a reorientation. "I must change to stay the same," he often said, as if suggesting that a change in his technical procedures allowed him to best express his evolving emotional response to his new surroundings (T. Hess 1959a, 13). During the 1960s de Kooning searched for a richly workable medium and a more painterly method of applying his paints. As the grit and grime of Manhattan was the prevailing theme of the pictures of the 1940s and 1950s, light, flesh, and water became the overriding subjects of the paintings executed in East Hampton. As a means to that end, the materials de Kooning used in the paintings of the 1960s—the supports, the pigments, and the binding media for his paints—all changed significantly.

This chapter draws on an in-depth technical examination of six paintings executed from 1962 to 1965. The paintings include *Untitled* (May 1962); *Two Standing Women* (1963–64); *Sphinx* (1964); *Woman* (1964); *Woman, Sag Harbor* (1964); and *Woman* (1965). The results of medium analysis of six additional paintings executed between 1960 and 1967—*Door to the River* (1960); *Spike's Folly II* (1960); *Pastorale* (1963); *Rosy-Fingered Dawn at Louse Point* (1963); *The Visit* (1966–67); and *Two Figures in a Landscape* (1967)—are also included.

Abstract Landscapes: *Untitled* (May 1962)

In May 1959, de Kooning exhibited a number of large oil paintings at the Sidney Janis Gallery in New York, which signaled a new direction in his work. Recognized as a departure from the densely packed urban themes of the previous years, the pictures were

FIGURE 29

Untitled, May 1962. Oil on canvas, 202.9 × 179.5 cm (79⅞ × 70⅝ in.). Hirshhorn Museum and Sculpture Garden, Smithsonian Institution (66.1224). Gift of the Joseph H. Hirshhorn Foundation, 1966.

© 2010 The Willem de Kooning Foundation / Artists Rights Society, New York. Photo: Lee Stalsworth

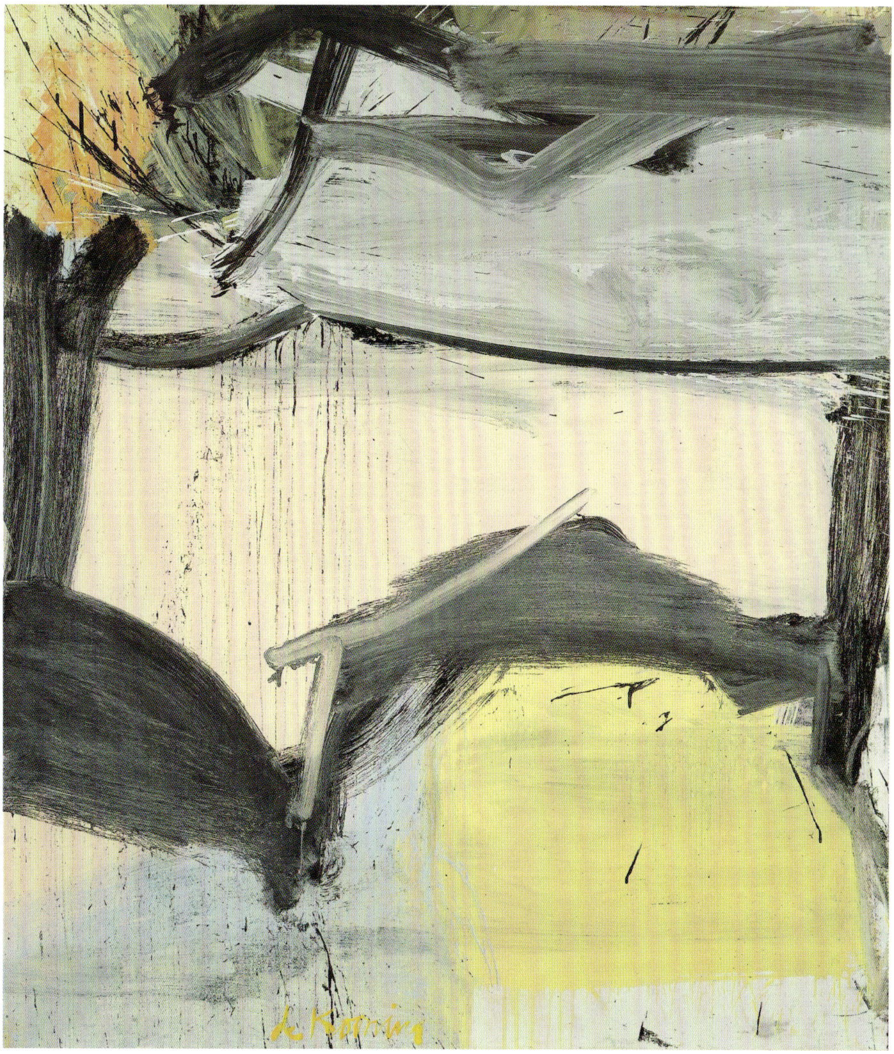

described as having a "huge landscape effect" (T. Hess 1959a, 13). These bold and broadly simplified abstractions, which suggest wide and radiant vistas, predated de Kooning's move to Long Island but were clearly painted from the perspective of the countryside. De Kooning told Irving Sandler in 1959 that in these new paintings he was trying to "grab a piece of nature and make it as real as it actually is" (Sandler 1989, 217).[66] He later told David Sylvester that he was working on a painting in his New York studio when he realized that it suggested the emotions associated with glimpsing the passing countryside from a moving car: "landscapes and highways and sensations of that, outside the city—with the feeling of going to the city or coming from it" (Sylvester 1963, 47–48).

Titles of several paintings of this period indicate the artist's shift in orientation. They often refer to sensations or emotions he associated with a particular location, season, or climate. Although the paintings are not descriptive, they suggest subliminally a horizon line, the heat and humidity of a late summer day, or a narrow strip of an asphalt parkway. Yet landscape associations and abstraction co-exist in a dynamic relationship. Even as the brushstrokes and colors suggest a particular locale, they simultaneously refer to the act of painting. As Andrew Forge noted, "Here the landscape sensations are very literal; they are also very unstable, for as one presses the pictures closely they have a trick of returning to nothing but paint" (Forge 1968, 248). To a much greater extent than in the previous paintings, the artist's emphatic gesture can be traced as he moved his brush from one edge of the painting to the other, often extending it beyond the frame of the picture. De Kooning's graphic, nonreferential treatment of paint was aptly described by Harold Rosenberg as the process by which "painting became its own subject" (Cummings 1983, 19).

From 1957 to 1963, de Kooning's large paintings usually measured 80 by 70 inches (203 × 178 cm). He favored this almost-square format, he said, because it made a big painting seem "intimate through appearing smaller than it is" (Rosenberg 1972, 56). He may also have favored this shape because he tended to turn a painting as he worked on it. De Kooning also explored this format on a smaller scale in a number of oils on paper and on fabric with similar proportions. Some of these were independent paintings and others were fragments cut from larger paintings, which in turn suggested ideas for still other larger works.

Untitled (May 1962; fig. 29) typifies the shift in de Kooning's pictorial conception during the late 1950s and early 1960s. Measuring nearly 7 by 6 feet, it was executed on cotton duck fabric, which the artist sized with a heavy layer of animal glue. The sizing was followed by a relatively thick, opaque white oil ground that largely masks the canvas weave. The pigment composition of the ground layer, a homogeneous mixture of titanium calcium white (TiO_2+$CaSO_4$) with small amounts of clay (aluminum silicate + iron oxide), indicates that the paint of the priming layer probably was a commercial paint that de Kooning purchased rather than one he prepared himself.[67] While the picture reveals evidence of considerable scraping, it did not evolve through the extensive process of revision that characterized de Kooning's previous work. Additionally, the paint lacks the gritty build-up and scabrous intermingling of enamel, charcoal, and granular particles that characterizes the paintings of the late 1940s and 1950s. Instead, the more fluid paint, in colors pared down to a few hues, was reduced to a few broad strokes. To a greater extent than in the previous paintings, the viewer can trace the artist's "housepainter's wide gesture," as he moved his brush edge to edge, often extending it beyond the frame of the painting (T. Hess 1959a, 13).

In *Untitled* (May 1962), broad horizontal and vertical vectors in gray-brown paint move across expanses of pale pink, yellow, and blue, echoing the frame edges. The spare geometry of the artist's brushwork is moderated by drips and spatters of paint at the ends of the brush marks. Clearly evident in the top left corner is the way in which the artist quickly twisted and fanned out the bristles at the end of his stroke as he lifted the brush from the painting surface, sending sprays of paint diagonally toward the corner. Years later, de Kooning explained that he painted these "so-called landscapes" with inexpensive fiber brushes normally used to paint ceilings. He conditioned the brushes by boiling them in water until they became "kind of rubbery" and would "hold the paint" (De Antonio and Tuchman 1984, 54). As before, de Kooning spent hours of study and contemplation while creating these images, but the actual execution was done quickly. As he explained, "I have to do it fast. It's not like poker, where you can build to a straight flush or something. It's like throwing dice" (*Time* 1959, 72). Although de Kooning freely acknowledged that he was an "action painter," he bemoaned the public's perception that "it's just like wild painting." Rather, he explained, "Lately I feel like a very talented house painter. . . . I have more paint now. And I can dig into it and I get excited once in a while and I maybe make a good painting, I hope" ("Soirée" 1959, 2).[68]

While de Kooning worked on small paintings during vacations on Long Island, the larger paintings of this series were executed in his new studio on lower Broadway, where he moved in 1958. A 1962 photograph taken by Rudolph Burckhardt of the new

FIGURE 30
De Kooning in his Broadway studio, New York, 1962, showing *Untitled* (May 1962) on the easel and glass bowls on the floor that the artist used to mix his paints.

Photo © 2010 Estate of Rudy Burckhardt / Artists Rights Society (ARS), New York. Artwork © 2010 The Willem de Kooning Foundation / Artists Rights Society, New York.

FIGURE 31
Detail from the lower right quadrant of *Untitled* (May 1962), showing the impressed weave pattern made when the artist pressed a piece of fabric against the still-soft paint. Popped bubbles in the paint surface indicate that the artist was adding water to his paints. Photographed in raking light.

studio shows *Untitled* (May 1962), propped against a portable easel (fig. 30). On the floor to the left of the painting are rows of bowls filled with paint. De Kooning told Harold Rosenberg that during this period he began making his own colors in order to incorporate the feeling of natural light in his painting. He also said that he worked out this system because the colors he selected remained wet longer and they did not dry unevenly (Rosenberg 1972, 58; Shirey 1967, 80; De Antonio and Tuchman 1984, 55). By the late 1950s, he apparently had begun to prepare his preferred color mixtures in quantity prior to beginning a painting. Starting with a wide range of basic colors, he would blend different hues as needed (T. Hess 1968, 103).[69]

Pigment analysis of the paints in *Untitled* (May 1962) (see Appendix 1, Table 2) reveals that most of the colors are not complex mixtures but instead are straightforward combinations of one or two colored pigments, often blended with large amounts of white—usually varying amounts of titanium dioxide, zinc oxide, and blanc fixe (barium sulfate)—to create pastel tones. The yellow paint is a cadmium yellow lithopone ($CdS+BaSO_4$) combined with zinc white. The pink is an organic red pigment mixed with titanium dioxide and small amounts of zinc oxide. The blue is a phthalocyanine blue combined with titanium white. The brown paint is a mixture of several cadmium-based colors with an organic blue and red dye. De Kooning's preparation of pale values in this work and others in the series from 1960 through 1963 forms a striking contrast to his earlier preference for the high-keyed hues that he called "circus colors." His selection and preparation of pastel colors appears to be a response to the sun-bleached colors of his new surroundings. With pigments mixed mostly with white, de Kooning suffused his colors with a pale luminous tonality evocative of the hazy light of the North Atlantic coast.

De Kooning's choice of a light-colored ground also plays a key role in imparting luminosity to the overlying paint, especially as the ground layer is often a component of the final design. Working from the basic color of the white ground, the artist laid in white-based mixtures of yellow, pink, and blue. These paints were sufficiently dilute in some areas that the lower white layer is partially visible in glimpses through the colored forms. In the top left corner, the ground remains unpainted, clearly visible as part of the final image. Just as the artist deliberately eschewed priming layers in many of his paintings of the late 1940s and 1950s to better enhance the coarse texture of his gritty paint, a smooth, white ground or light-colored support became characteristic of his work of the 1960s. The white underlying tone was clearly premeditated, well-suited to the atmospheric effects that he sought to capture. One senses, rather than identifies, the sky and water in the pale colors. Years later when he was shown reproductions of these landscape paintings, de Kooning commented on their pale luminous effects, saying that they were "just like you see only the sky" (Wolfe 1981, 13).

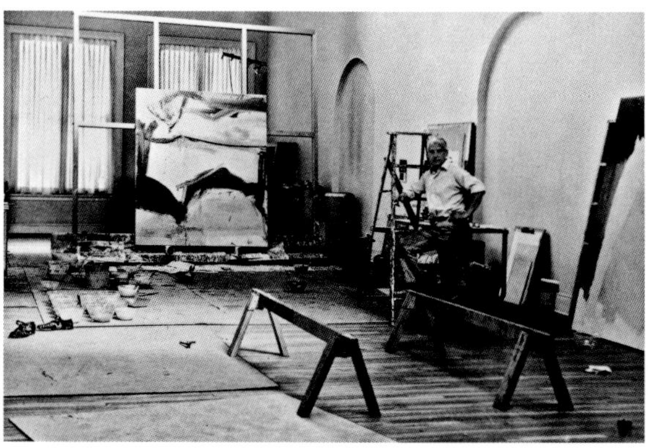

Examination and technical analysis of *Untitled* (May 1962) and other paintings in this series indicate that de Kooning was continuing past practices but was also experimenting with methods to keep his paints fluid for longer periods of time. By 1955, his paintings had begun to sell, and, as mentioned earlier, his dealers were willing to advance him money to buy art materials. For the first time in his life he could purchase enough quality art supplies to suit his needs (T. Hess 1968, 100). Despite his increased affluence, he continued to use house paints, though on a more limited basis. The dark gray-brown paint in *Untitled* (May 1962) is an alkyd-based house paint, as is the light gray paint in *Pastorale* (1963).[70] Yet de Kooning's art seems increasingly to have required properties that neither traditional nor commercial retail trade paints could offer. While continuing to use these fast-drying house paints, he simultaneously began to experiment with methods of slowing the drying rate of his paints. One of de Kooning's problems since the late 1940s had been how to keep his paints malleable during the extended time he worked on his pictures. Artist Landès Lewitin once recommended oil of cloves as an old trade secret (Gaugh 1983, 61). Possibly as early as the mid-1940s and surely by the 1950s, de Kooning began to press paper, usually newspaper, against sections of his painting surfaces overnight to slow the drying process and to keep the paint from becoming unworkably thick (Prather 1994, 134).[71] He also pressed fabric against his wet paints. A distinctive reverse weave pattern seen in the surface of the yellow paints of *Untitled* (May 1962) clearly resulted when a piece of cloth was pressed against the paint while it was still soft (fig. 31).

Medium analysis indicates that de Kooning also attempted to retard the drying rate by adding poppy-seed oil (or poppy oil) to some of his artists' oil paints. The P/S peak area ratios of samples of the pink and yellow paints from *Untitled* (May 1962) are 4.8 and 4.4, respectively (see Appendix 1, Table 2), indicating poppy oil or possibly a mixture of poppy oil and linseed oil as the binding medium. P/S molar ratios as well as azelate to glycerol (A/G) and palmitate to glycerol (P/G) molar ratios of paint samples of the light blue paint of *Pastorale* (1963), the off-white and pink paints of *Door to the River* (1960), and the blue paint of *Spike's Folly II* (1960) also correlate well with poppy oil or a mixture of poppy and linseed oils (Schilling and Khanjian 1996, 220–27).

Poppy oil is rarely used in retail trade paints due to its high price and its slow drying rate (Heaton 1947, 235). Rather, it is principally used as a binding medium in artists' oil colors, especially with light-colored pigments, due to its pale color and its tendency to yellow less than linseed oil upon aging. Its slowness of drying has long been recognized among artists as an advantage; it permits alterations for a considerable time after the paint has been applied. De Kooning may have consciously selected artists' tube paints with a binding medium of poppy oil because he was aware of this oil's slower drying rate. More likely, he mixed unpigmented poppy oil with his artists' tube paints. The analytical results are supported by a casual remark de Kooning made to an interviewer in 1969, when he acknowledged that he had once experimented with adding poppy oil to his paints but eventually gave it up because it was too "oily" (Perreault 1969, 47).

In *Untitled* (May 1962) and other works of this period, traces of tiny, popped bubbles on the surfaces of the paint film offer evidence that de Kooning was also mixing water with his paints (see fig. 31). Reportedly, the artist began to mix a few drops of water with his oil paints, usually in salad bowls, in the late 1950s (Gaugh 1983, 120). While he may have added water in the hope that it would slow the drying rate of his paints, it is also likely that he was attempting to capture the spumy appearance of his watery environment by literally whipping water in his paint medium.

The late 1950s and early 1960s marked a period of stylistic and technical transition for de Kooning, which *Untitled* (May 1962) clearly reflects. In his 1959 interview with Irving Sandler, the artist admitted that he was "grappling for a way to say something new" and that he had "this point of reference—my environment—that I have to do something about" (Sandler 1989, 216). His use of a smooth, opaque white ground, his admixture of large amounts of white to his paints to give them a pale luminosity, and the alteration of his binding medium by adding poppy oil and water illustrate the degree to which de Kooning was engaged in a substantial change in his method of painting. As he was preparing to move from Manhattan to Long Island, he apparently was adapting his procedures and his choice of materials to accord with his new subject matter.

Return to the Figure: *Two Standing Women* (1963–64), *Woman* (1964), and *Sphinx* (1964)

After living and working in Manhattan for nearly forty years, de Kooning began in late 1962 to move from his studio to Springs, Long Island, and by mid-1963 he was living there year-round. After months of involvement with the design and construction of his new studio, he returned to a renewed examination of the human figure. The female figures in his work of the 1960s are substantially different from those of the late 1940s and 1950s. De Kooning acknowledged that these new women, while marking a return to an old subject in his art, were of an altered spirit. "The women I paint now are very friendly and pastoral, like my landscapes, and not so aggressive," he said (Shirey 1967, 80).[72] He also expressed his growing confidence with the act of painting. "I'm going back to what I used to do," he said, "but now I can do it better. I feel competent and yet still in enough of a huddle to feel excited" (Dickerson 1964b, 71).[73] Examination and technical analysis of *Two Standing Women* (1963–64), *Woman* (1964), and *Sphinx* (1964) reveal that these are pivotal paintings at this point in the artist's career. With these paintings he tested procedures that he would exploit further throughout the remainder of the decade. In the first, he experimented with ways to render the impression of the light and water in his new environment. In the second, he manipulated his binding medium and methods of applying his paints to better convey the sensuality of human flesh. In the third, he experimented with the method of oil transfer and adding more water to his paints.

A group of small nudes on paper mark de Kooning's earliest attempt in this new series of women. These paintings, some of de Kooning's most naturalistic depictions of women, were based on actual figures seen wading in the water, walking along the shoreline, or sitting on the beach. In *Two Standing Women* (fig. 32), two nude figures wade in shallow water along the beach, their silhouettes and reflections wavering and shimmering in the reflected light and heat. In 1964, de Kooning told an interviewer that he was working on a water series: "The figures are floating, like a reflection in the water. The color is influenced by the natural light. That's what is so good here" (Willard 1964, 44). A few years later he told Daniel Frasnay that he was trying to capture in these works the ephemeral images he glimpsed of waders on the beach. "Everything should float. When I go down to the water's edge... I see the clam diggers bending over, up to their ankles in surf, their shadows quite unreal, as if floating. This is what gave me the idea" (Frasnay 1969, 221).

As with *Untitled* of May 1962, the impression of strong light in *Two Standing Women*, like the paintings that followed, is due in large part to the artist's selection and preparation of his supports. This work was executed on a piece of white Bristol board, a machine-made paper with a surface that is heavily sized and polished.[74] Thomas Hess

FIGURE 32
Two Standing Women, 1963–64. Oil and charcoal on paper, mounted on canvas, 73.7 × 58.8 cm (29⅛ × 23¼ in.). Hirshhorn Museum and Sculpture Garden, Smithsonian Institution (66.1206). Gift of Joseph H. Hirshhorn, 1966.

© 2010 The Willem de Kooning Foundation / Artists Rights Society, New York. Photo: Lee Stalsworth

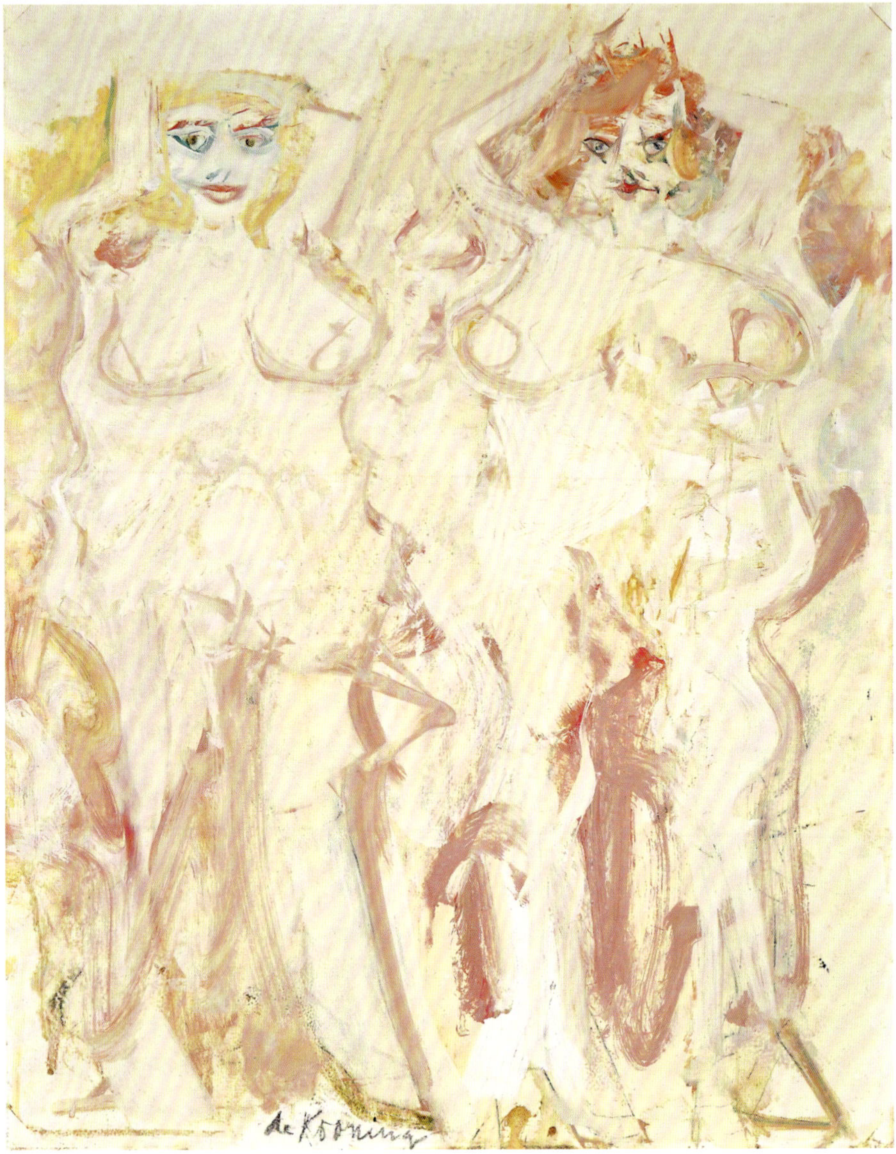

frequently mentioned de Kooning's lifelong preference for such slick, smooth surfaces, which, he said, enabled the artist to paint with increased speed (T. Hess 1972c, 15). The artist's partiality for supports with smooth finishes, in fact, seems to have coincided with his extended visits to and eventual move to Long Island. During the late 1940s and throughout most of the 1950s, as has been shown, de Kooning selected his supports for their coarse textures and he generally left them unprimed. After about 1959, he selected and prepared his supports to accentuate their hard, slick surfaces. Translucent machine-made "vegetable" vellum papers with heavily sized and calendered surfaces and fabric and wood panels with oil-based ground layers, sanded to a smooth finish, are all evident in paintings of the 1960s.[75] These carefully prepared supports were clearly a conscious choice that not only allowed the artist to apply his paint with little resistance or drag, but also provided nonabsorbant surfaces that permitted him to accentuate his medium-rich paint.

De Kooning reportedly did not paint much during the last months of 1963 and into 1964. Rather, he obsessively worked and reworked *Two Standing Women*. According

FIGURE 33
Detail from the right edge of *Two Standing Women*, showing residues of scraped and sanded preliminary paint layers and the "orange-peel" pattern of the clear resin that the artist applied to the support prior to the application of paint. Photographed in specular light.

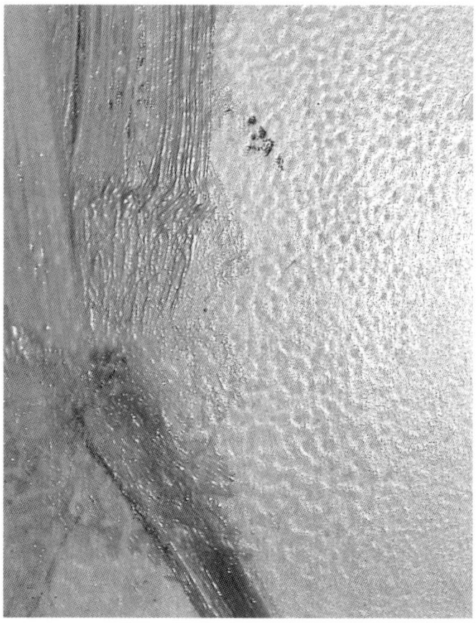

FIGURE 34
Untitled, 1966. Charcoal on paper, 43.1 × 35.6 cm (17 × 14 1/16 in.). Hirshhorn Museum and Sculpture Garden, Smithsonian Institution (86.1345). The Joseph H. Hirshhorn Bequest, 1981.

© 2010 The Willem de Kooning Foundation / Artists Rights Society, New York. Photo: Lee Stalsworth.

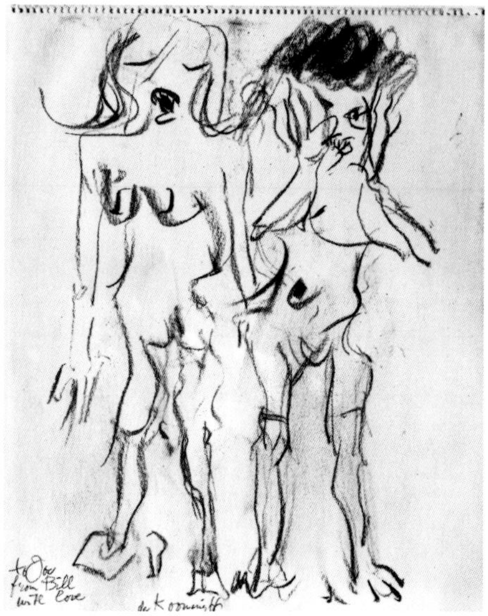

to Susan Brockman, de Kooning's girlfriend at the time, this work is a transitional painting. Brockman's assertion is supported by technical examination (Stevens and Swan 2004, 457).[76] Although the final image of *Two Standing Women* undoubtedly was executed quickly, the work itself evolved through several stages. The first stage was a cursory sketch in graphite drawn directly on the paper support. In this drawing de Kooning delineated only the women's heads, lower legs, and feet and placed the two women symmetrically and parallel to the edges of the white paper, with no landscape setting to anchor their buoyant forms. He next applied a heavy, irregular coat of an unpigmented resin that covers the drawing and indeed the entire surface of the paper. He then applied several layers of pastel-toned paints of various hues, which he then repeatedly scraped or sanded down so that only residues of the different colored paints can be seen as a stippled pattern around the women's forms. Finally, the artist skillfully outlined the women's fluid forms with softly looping brushstrokes, following the basic guidelines of the drawing. The women's almost indeterminate shapes in pale, low-contrast pinks, blues, and yellows, placed against the all-white areas of exposed paper—seen around and between the figures—suggest a bright outdoor light that lends the figures much of their buoyancy.

The flickering, *plein air* atmosphere of *Two Standing Women* is also in large part due to de Kooning's experimental use of an irregular resin layer, placed between the support and the paint layers. Identified as a methacrylate resin, the acrylic probably was applied as a spray coat. In places where it was applied most heavily, it has assumed a dimpled "orange-peel" or "wave-formation" pattern (fig. 33).[77] This acrylic resin layer serves several functions, all which center upon ways to convey light. First, the highly reflective surface of the coating, which covers the entire surface of the paper, produces a sense of refracted, shimmering light, especially where it is left exposed. Second, the resin serves as an isolating layer that separates the paint from the support. Consequently, the paint could not sink into an absorbent ground or support. Rather, it pooled above the paper, suspended on the transparent resin. Thus, light is refracted through the clear resin film and is reflected from the paper beneath it, in the same way that a reflection appears to be suspended on water. Third, residues of the initial scraped and sanded paint caught in the recesses of the resin were certainly deliberately left by the artist to conjure an impression of dappled light bouncing off of water. Finally, the paint strokes

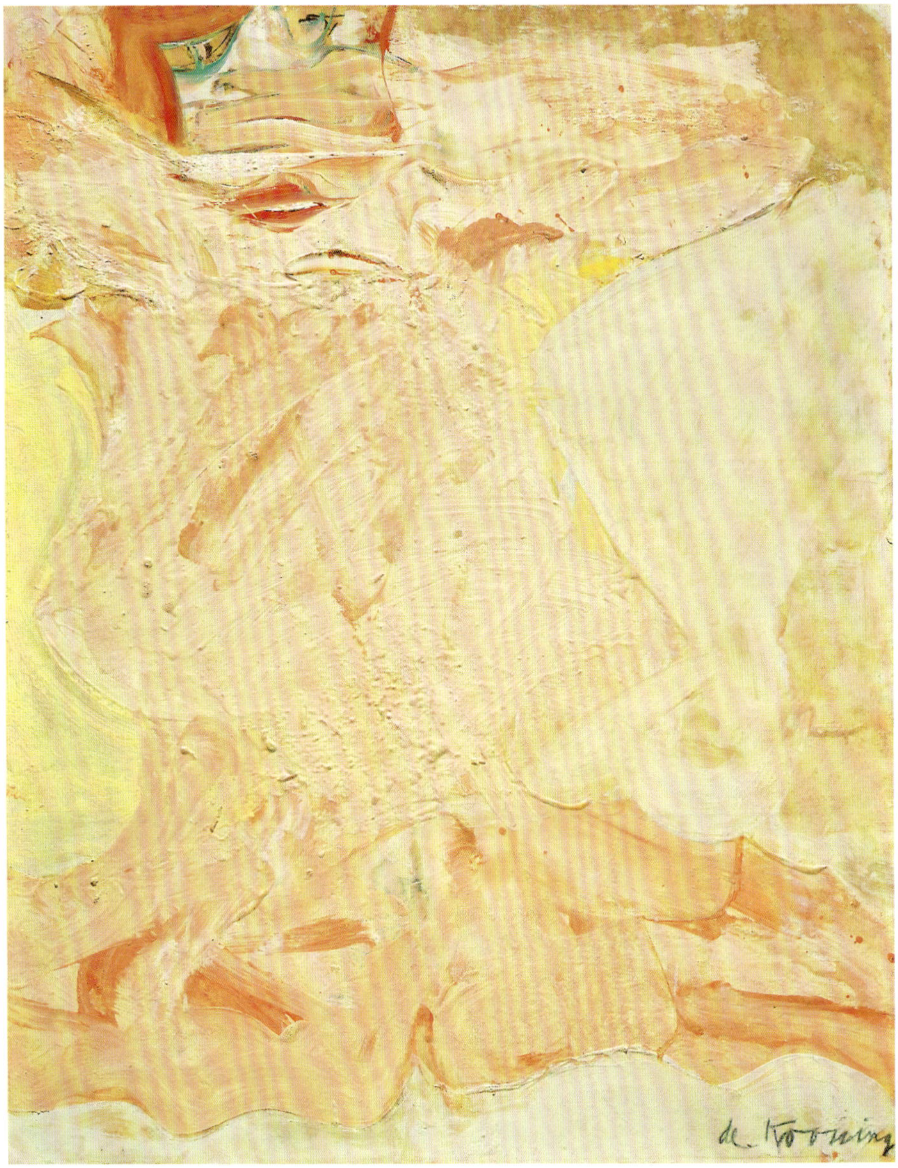

applied to the undulating resin have conformed to the irregular surface, and these wavy lines have assumed a soft, rippling quality that imbues the female forms with much of their fluidity, as if they are partially dematerialized by water and shimmering light. This imagery is also evident in drawings of the period (fig. 34). Within a short time, de Kooning would begin to manipulate his binding medium so that the paint itself would assume an irregular, puckered surface. In *Two Standing Women*, apparently an isolated experiment, he used one of the newly developed acrylic resins as a means to present an image that appears to ripple across a watery surface.[78]

Two Standing Women is one of the most diffuse images de Kooning produced. Pale, low-contrast colors placed against an all-white background and a reflective undulating acrylic resin result in an image startling in its fidelity to natural observation. The images of these bawdy women have often been discussed in terms of the bleached light of Long Island, but they have also been seen as evoking a Baroque conception of beauty, a comparison de Kooning did not deny: "Yes, maybe they do look like Rubens.

FIGURE 35
Woman, 1964. Oil and charcoal, mounted on fiberboard, 73.7 × 58.4 cm (29 × 23 in.). Hirshhorn Museum and Sculpture Garden, Smithsonian Institution (66.1212). Gift of Joseph H. Hirshhorn, 1966.

© 2010 The Willem de Kooning Foundation / Artists Rights Society, New York. Photo: Lee Stalsworth.

FIGURE 36
Woman, 1964, photographed in raking light.

Artwork © 2010 The Willem de Kooning Foundation / Artists Rights Society, New York

Yes, Rubens, with all those dimples" (Willard 1964, 44). Throughout the remainder of the decade, the artist would continue overtly to explore the sexual aspects of his female figures.

The altered mood of de Kooning's new women is obvious in *Woman* (1964; fig. 35). In this work, he explored eroticism as a subject not only in terms of the subject presented but also by means of his evolving techniques. Rather than the more discreet reclining pose common in Western art, the artist placed his nude frontally, squatting in a lascivious pose, with legs splayed and arms raised to cover her face. Thomas Hess likened the position, used repeatedly throughout the 1960s in paintings and drawings, to that in Ingres's portrait of Madame Bertin (Musée du Louvre, Paris; T. Hess 1967, 34). Cooke argued that the pose is so provocative that it belongs to the age of *Playboy* and *Hustler* magazines (Cooke 1993, 95). The pose undoubtedly was also an extrapolation of de Kooning's observations of everyday life: sunbathers on the beach who lay sprawled on bathing towels or seated in lawn chairs.

De Kooning began *Woman* (1964) with a cursory charcoal drawing on bristol board, visible in infrared photographs, that delineates only the woman's eyes, arms, and genitalia. As with *Two Standing Women*, the artist applied a layer of transparent resin to the paper support, covering the drawing and probably the entire support.[79] In this work the resin layer is very thin and almost entirely covered by the upper paint layers. Thus, de Kooning appears to have applied it more to provide a slick smooth surface on which he could spread and manipulate his increasingly fluid paint medium, rather than as a means to convey bright, reflected light.

Medium analysis of paints in this picture and those in *Two Standing Women* indicates that de Kooning was continuing to add poppy oil to his tube paints. The P/S peak area ratios of samples of the yellow and light pink paints of *Two Standing Woman* are 4.3 and 4.1, respectively, and the P/S ratios of the pink and yellow paints of *Woman* (1964) are 4.3 and 4.5, respectively, all clearly indicating poppy oil (see Appendix 1, Table 2). Using this slower-drying medium, de Kooning could apply and work his paint over longer time periods and then scrape and wipe parts, or all, of the image down to a

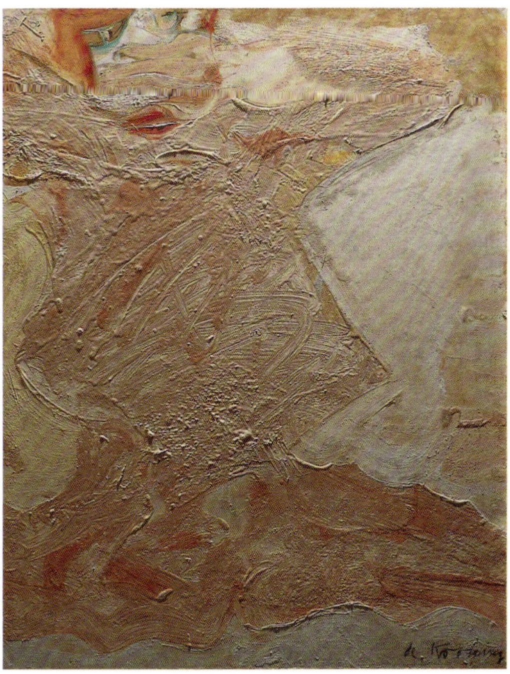

trace. The artist's working method is graphically apparent in raking light photographs of *Woman* (1964; fig. 36). He depicted the woman's almost formless body with very few brushstrokes and without any interior delineation.[80] All volume is provided by a generous build-up of heavily worked pink paint. Readily evident are the tracks of the brush, and possibly its handle, where the artist stirred and mixed different pinks on the smooth paper support just as the paint had begun to form a skin. His use of a scraper or palette knife is also apparent in the raking light photograph. Clearly visible is the way the artist ran the straight edge of the blade around the perimeter of the woman's form, scraping the yellow and white paints almost down to the paper support and leaving her pink torso, legs, and buttocks in almost sculptural relief.

While de Kooning may have mixed poppy oil with tube paints in order to work his paints for longer periods of time, he was also doing so to achieve certain visual effects. He clearly was seeking a way of using paint to create an analogue to human flesh. *Woman* (1964) is a remarkable exercise in the sensuous materiality of paint. With the woman's torso nearly filling the frame and the specific details of her form subsumed in an increasing painterliness, the viewer is drawn close to the picture. From this vantage point, medium-rich paint suggests a lush, unctuous material that inevitably reminds one of wrinkled, puckered flesh. In this image and in others of women painted during the 1960s, there is "an extraordinarily powerful suggestion of moving flesh, which responds to the brushstroke as if to the touch of a caressing hand" (Cooke 1993, 103). As in the paintings of the 1940s, de Kooning was not content to illustrate just the visual appearance of a form. His paints of the 1960s, often described by both admirers and detractors as "juicy," "succulent," and "sleazy," and the textures peculiar to those paints, characterized as "puckered," "wrinkled," and "flaccid," are metaphors for human flesh. As such, they are evocative of sensations that de Kooning, who by his own admission greatly admired and enjoyed women, had experienced directly.[81]

While sexual implications are present in virtually every phase of de Kooning's work, that interest is more readily evident in the works of the 1960s.[82] The overtly erotic pictures of the 1960s are undoubtedly related, in part, to the artist's long-standing interest in the calendar nude, images in advertising, and his move to a beach community. They can also be placed within the social context of the 1960s, with its unprecedented emphasis on sexuality.[83]

Sphinx (1964; fig. 37) exemplifies de Kooning's experimentation with the process of oil transfer and with adding water in his oil medium. The image, a man posed doing a handstand, resulted when de Kooning pressed a sheet of vegetable vellum against still-wet paint that had been applied directly to an unprimed wooden surface. When the paper was removed, it presented a reverse image of the painting, or a portion of the painting, at a certain point in its evolution.[84] The oil transfers, on newspaper and vellum, grew out of de Kooning's practice of pressing newspaper sheets against the wet paint surface of a work in progress to prolong the drying time of his paints. The artist initially left these transfers scattered around the studio to document and study the stages through which a painting had gone and to suggest new ideas (fig. 38). Later he considered them independent works worthy of exhibition and sale.[85] Sometimes the artist reworked the transfer images, painting directly on the offset sheet. Occasionally, he traced an image in charcoal while the paper sheet was still pressed against the painting. Once the sheet was removed, both the paint and the charcoal remained visible as part of the final image.[86] In the case of *Sphinx*, the inverted image was an association de Kooning apparently made after the transfer was completed. The eyes in black paint are part of the transfer but the charcoal lines delineating the figure's arms and hands were added only after the paint had dried completely.

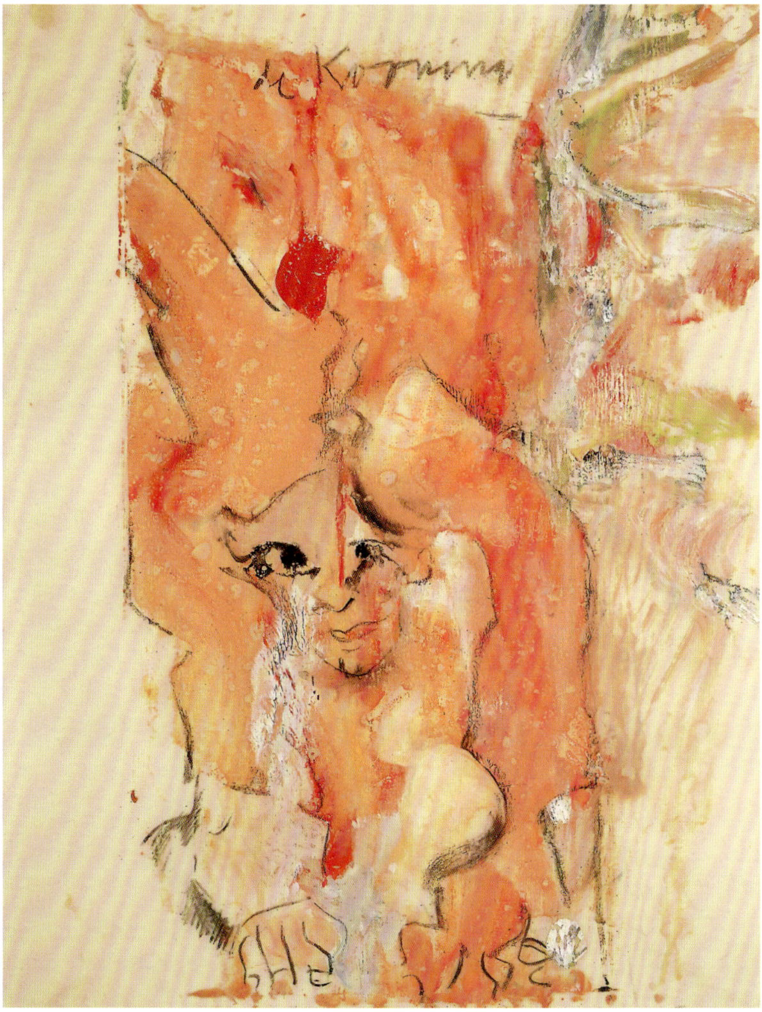

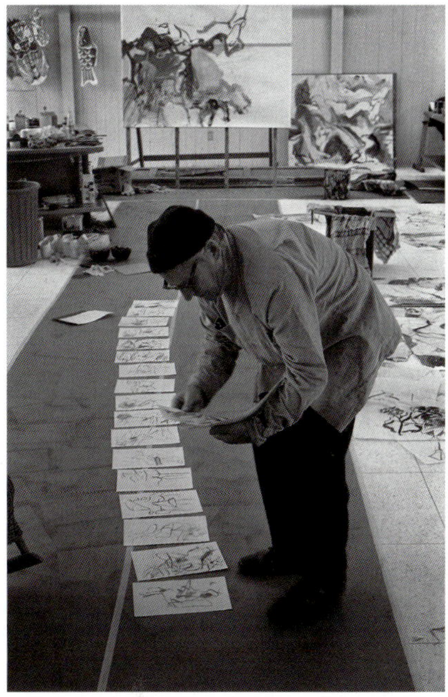

FIGURE 37

Sphinx, 1964. Oil and charcoal on paper, 60.5 × 47.3 cm (23 13/16 × 18 5/8 in.). Hirshhorn Museum and Sculpture Garden, Smithsonian Institution (66.1201). Gift of Joseph H. Hirshhorn, 1966.

© 2010 The Willem de Kooning Foundation / Artists Rights Society, New York. Photo: Lee Stalsworth

FIGURE 38

De Kooning in his studio, Springs, East Hampton, 1971. "Oil transfers" on the floor behind him resulted when the artist pressed sheets of paper against the surfaces of paintings in progress.

Artwork © 2010 The Willem de Kooning Foundation / Artists Rights Society, New York. Photo: © Dan Budnik

The process of oil transfer also allowed de Kooning to experiment with the intriguing visual effects made possible by the natural antipathy of oil and water. Large, round voids in the paint surfaces of *Sphinx* and other oil transfers on vellum of the period resulted when water and air bubbles caught in the paint matrix burst during the transfer process. These voids are readily visible throughout the image and are particularly apparent in an ultraviolet photograph (fig. 39). De Kooning may have created this effect by adding greater percentages of water to his paint medium. He also may have brushed water onto the paint surface or to the vellum sheet before he pressed the two surfaces together. In either case, the final image is similar to several of the lithographic images de Kooning produced from 1960 through 1971 (fig. 40). It has been suggested that de Kooning's experience with lithography, beginning in 1960 and revisited in 1966, may have initiated his experimentation with the more liquid binding medium evident in the paintings of the late 1960s and the 1970s (Cooke 1993, 109).[87] As has been shown, de Kooning was already adding water and larger amounts of oil to his paints as early as 1959. It is intriguing, therefore, to consider that the oil transfers may have sparked the artist's interest in lithography, which in turn stimulated his interest in increasingly fluid paints made with the addition of still greater percentages of oil and water.

Two Standing Women (1963–64), *Woman* (1964), and *Sphinx* (1964) exemplify de Kooning's exploration of technical methods that he developed throughout the

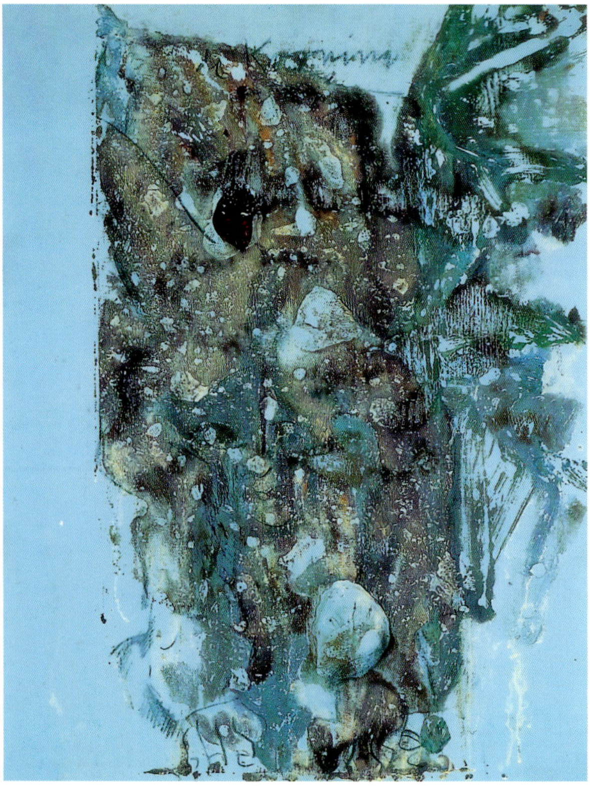

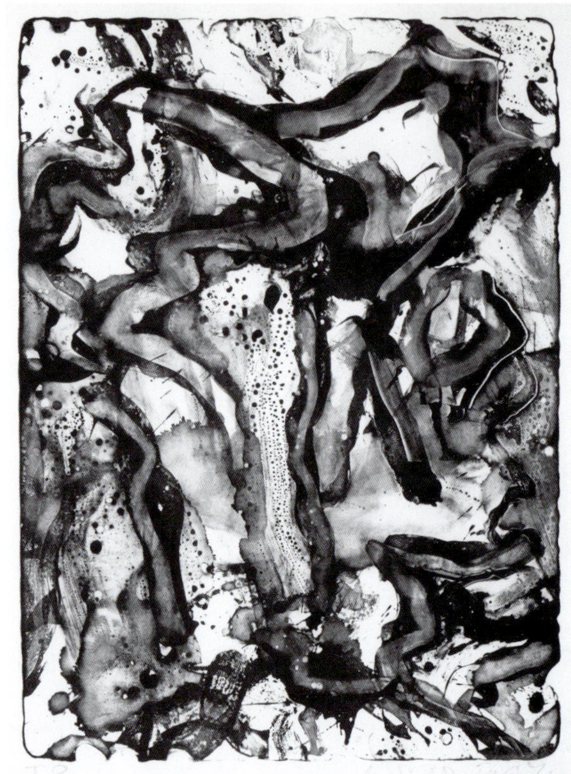

FIGURE 39
Sphinx, photographed in ultraviolet light.

Artwork © 2010 The Willem de Kooning Foundation / Artists Rights Society, New York

FIGURE 40
Landscape at Stanton Street, 1971.
Lithograph drawn on stone.
Black ink on Dutch etching paper.
76.2 × 55.9 cm (30 × 22 in.).
Published by Hollanders Workshop and Fourcade, New York, 1971.

Artwork © 2010 The Willem de Kooning Foundation / Artists Rights Society, New York

remainder of the decade and into the 1970s. His procedures may be seen as experimental and even fortuitous. Yet many of his ideas concerning the preparation of his supports, the mixing of artists' paints with increasing amounts of oil and water, his exploitation of oil transfer, his handling of paints, and the use of color to exploit impressions of outdoor light, water, and flesh, were already present in these works and would become increasingly significant factors in his evolving methods.

The Door Paintings: *Woman, Sag Harbor* (1964), *Woman* (1964), and *Woman* (1965)

From 1964 to 1966, de Kooning painted a series of larger-than-life images of females on hollow-core doors. These works, considered a benchmark in the artist's career, confirm his interest in technical experimentation. With these so-called door paintings, including *Woman, Sag Harbor*, painted in 1964 (fig. 41), and *Woman* (1965; fig. 42), de Kooning was continuing his use of smooth, absorbent supports. He also persisted in mixing large amounts of white into many of his paints to give them a high-keyed luminosity. With a full-fledged facility in paint handling, he tested the potential of an increasingly fluid paint medium. As before, formal resolutions are indissolubly tied to technical innovations.

The door paintings resulted more or less accidentally from de Kooning's construction of a new studio on Long Island. The artist ordered several doors, measuring 80 by 36 inches (203 × 91 cm), from a local lumberyard. Once they arrived, however, he was unhappy to find that they were hollow-core rather than solid wood. When the supplier refused to take them back, de Kooning set them aside, propping them against

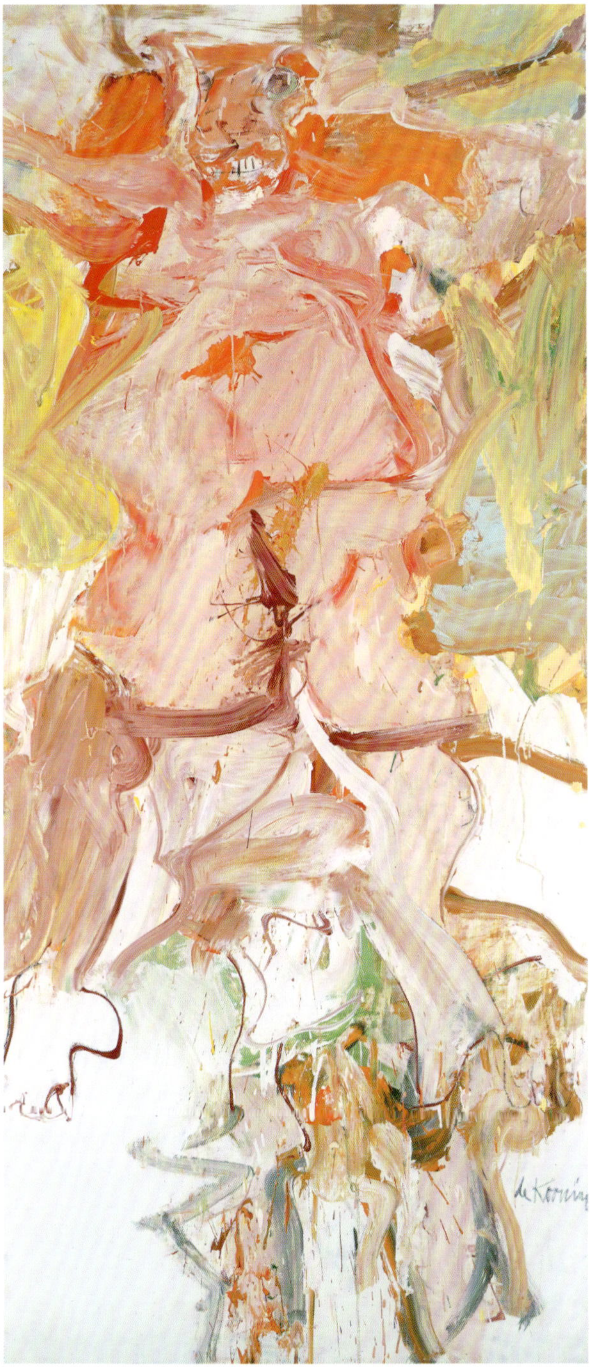

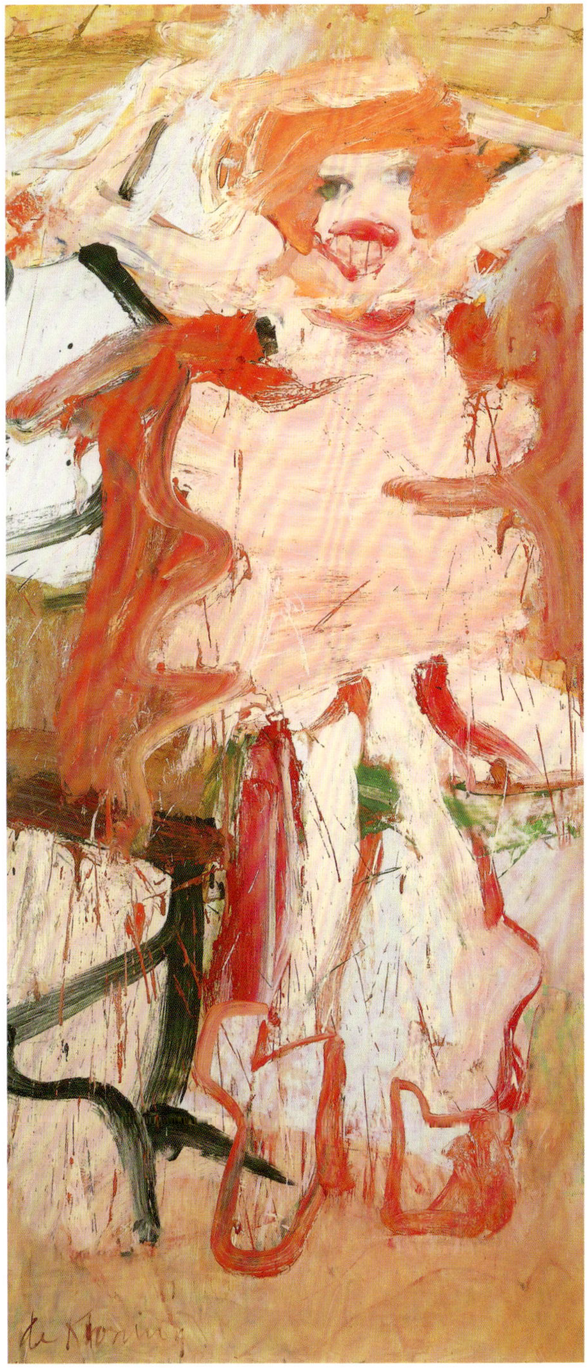

FIGURE 41
Woman, Sag Harbor, 1964. Oil and charcoal on hollow-core wood door, 203.1 × 91.2 cm (80 × 36 in.). Hirshhorn Museum and Sculpture Garden, Smithsonian Institution (66.1209). Gift of Joseph H. Hirshhorn, 1966.

© 2010 The Willem de Kooning Foundation / Artists Rights Society, New York. Photo: Lee Stalsworth

FIGURE 42
Woman, 1965. Oil on hollow-core wood door, 203.2 × 91.4 cm (80 × 36 in.), Hirshhorn Museum and Sculpture Garden, Smithsonian Institution (66.1217). Gift of Joseph H. Hirshhorn, 1966.

© 2010 The Willem de Kooning Foundation / Artists Rights Society, New York. Photo: Lee Stalsworth the studio wall. Later, while de Kooning was waiting for his studio assistant to prepare a series of canvases, he became attracted to the hard smooth surface of the doors. He debated using them as supports for paintings (McMahon 1993).[88]

The doors were eventually primed on the front and reverse with several coats of a white oil paint with a pigment composition of approximately equal parts titanium dioxide and zinc oxide with small amounts of kaolin (an aluminosilicate). John McMahon described the artist's elaborate specifications for preparing his grounds: up to six layers of priming were applied, with each layer sanded before the next was laid on, and the final coat was sanded to a smooth surface (McMahon 1993). Examination with a microscope and paint cross-sections from *Woman, Sag Harbor* and *Woman* (1965), however, reveals rather thin white grounds in a single layer. Examination of *Woman* (1965), in particular, shows only residual ground material, and along the bottom edge of the painting the wood grain of the door is readily visible through some areas of thinner paint. As both paintings show evidence of considerable scraping and sanding, it is likely that the artist abraded the multiple ground layers in the process of applying and removing his paints. Tom Ferrara, de Kooning's assistant during the 1980s, remembers that partially finished or abandoned paintings were regularly laid flat on sawhorses, and their surfaces were sanded with turpentine and water before painting was resumed (McMahon 1993; Garrels 1995, 22; and Kertess 2000).[89] Abraded areas in *Woman, Sag Harbor* and *Woman* (1965) indicate that this practice was common in the 1960s as well. De Kooning told art historian Judith Wolfe that he liked the effect of sanded and scraped areas in the final painting. From these residues, he culled images that suggested new ways of proceeding with the painting (Wolfe 1981, 16).

Woman, Sag Harbor, begun in June 1964 and finished by early fall of the same year, was the first of the door paintings (Fox 1995). As with earlier works, this picture is evocative of associations de Kooning made between a specific place and a general type of woman he observed:

> When I made *Woman, Sag Harbor*, I just titled it that, because I go to Sag Harbor and I like the town and I frequently have this association. While I was painting it, I said to myself, That's really like a woman of Sag Harbor, or Montauk, where it's very open and barren (Rosenberg 1972, 58).[90]

Through a series of complex maneuvers that included tracing and oil transfer, the artist transposed fragments of one painting onto the surfaces of others. *Woman, Sag Harbor* was the origin and catalyst of a number of other works, including *Woman* (1964; fig. 43) and possibly *Woman Accabonac* (1966; Whitney Museum of American Art), both made from early states of this painting (T. Hess 1965, 37).[91]

To create *Woman* (1964), de Kooning used a large sheet of vellum paper that already had a drawing in black paint along the lower portion of the paper. As he pressed the paper against the still-wet paint of *Woman, Sag Harbor*, he shifted the vellum slightly prior to pulling it off; this had the effect of blurring the transfer image and leaving a squiggly wave pattern in the paint. A comparison of the drawing elements of *Woman, Sag Harbor* with those in *Woman* (1964) reveals that the women's feet and shoes are near mirror images of each other. Thus, the drawing of a pair of woman's shoes with ankle straps in *Woman* (1964) is probably a transfer from an early version of *Woman, Sag Harbor*. This also indicates that an earlier version of *Woman, Sag Harbor* was conceived with the women straight-legged, wearing high-heeled shoes with ankle straps. The outlines of these earlier legs are still faintly visible, but at some stage the artist scraped down and painted over portions of the woman's parallel legs, adding two additional legs

FIGURE 43

Woman, 1964. Oil on vellum paper mounted on paperboard, 193.8 × 96.4 cm (76 3/8 × 35 3/8 in.). Hirshhorn Museum and Sculpture Garden, Smithsonian Institution (66.1208). Gift of Joseph H. Hirshhorn, 1966.

© 2010 The Willem de Kooning Foundation / Artists Rights Society, New York. Photo: Lee Stalsworth

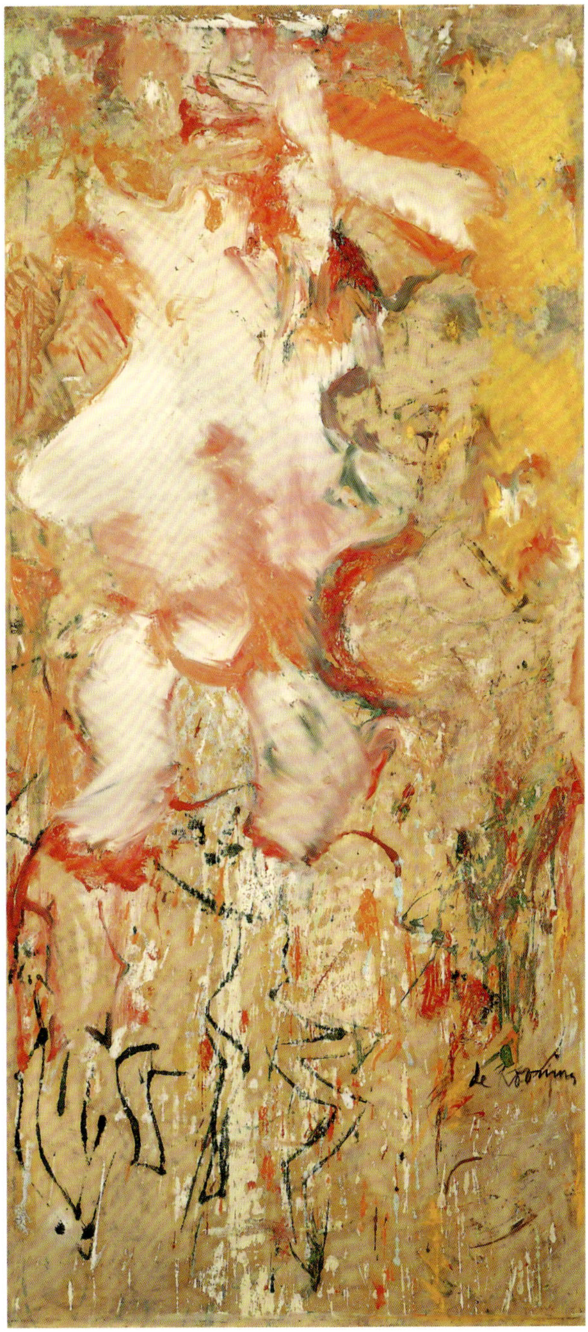

outlined in charcoal and maroon paint. One is a wavy high-heeled stump and the other is shoeless and gives a high kick. With as many as four legs and feet, the woman now appears to be in motion, jumping, dancing, or treading water.[92]

De Kooning probably employed other methods to generate the images of *Woman, Sag Harbor* and *Woman* (1965). In addition to his customary practice, described earlier, of using superimposed drawing fragments in the inception and development of a painting, the artist kept a large selection of drawings and photographs and recently completed paintings in the studio as resources. He often referred to a painting or drawing propped against or tacked to the wall near the easel, adopting lines from forms in these adjacent images as a way of making a drawing on a new canvas or accentuating lines on a painting in progress. Sometimes he would hold a drawing in his hand, sketching directly from it onto the canvas. Photographs of the artist working in his studio indicate that de Kooning continued to place fragments of drawings on the surfaces of paintings in progress. Occasionally he would tack or tape a full-sized charcoal drawing to the surface of a newly prepared support, with the charcoal lines facing the painting. He would then trace the drawing from the reverse, effectively transferring the charcoal lines to the support. John McMahon (1993) recalled that de Kooning would draw freely from any one of these sources, singly or in combination, to begin or proceed with a painting. Through a series of elaborate procedures, one painting or drawing brought forth many others, but the evolutionary process was seldom linear.

Color plays a significant expressive role in the door paintings. As mentioned earlier, when de Kooning first moved to East Hampton, he was so moved by the light and the colors of his new location that he changed his paints, mixing a range of closely related hues that better reflected the out-of-doors: "I felt in New York that I was using colors just prismatically: yellow, blue, black, white. I had no way of getting hold of the tone of the light of painting" (De Antonio and Tuchman 1984, 55–56). In 1972, he told Harold Rosenberg that he began making his own colors in order to incorporate the feeling of the natural light quite literally in his paintings. As he explained:

> When I came here I made the color of sand—a big pot of paint that was the color of sand. As if I picked up sand and mixed it. And the grey-green grass, the beach grass, and the ocean was all kind of steely grey most of the time. When the light hits the ocean there is kind of a grey light on the water.... Indescribable tones, almost. I started working with them and insisted that they would give me the kind of light I wanted (Rosenberg 1972, 58).[93]

As with *Untitled* (May 1962), identification of the pigments in *Woman, Sag Harbor*, *Woman* (1965), and *Two Standing Women* reveals a rather restricted palette of single colors and their tints, most of them modern pigments available only since the twentieth century (see Appendix 1, Table 2).[94] The intensely colored reds, oranges, and maroons in each of these paintings are cadmium reds and oranges, often in combination with organic pigments.[95] These pure pigments are used relatively sparingly in both pictures, and where they occur they have high chromatic brilliance. The paler pinks, oranges, and yellows of both pictures are cadmium colors mixed with various amounts of titanium dioxide, zinc oxide, and barium sulfate. The more white he added, the subtler the contrasts became. This narrow range of colors, used now without the black paints and duller earth pigments that characterized his earlier work, creates chromatic effects that are suffused with a pale luminous quality described by de Kooning as a "dazed kind of light" (Rosenberg 1972, 58).[96]

Photographs and anecdotal accounts suggest that by the time de Kooning painted *Woman, Sag Harbor* and *Woman* (1965), he had discontinued his use of house paints and poppy oil, instead adding safflower cooking oil to artist oil paints.[97] Dan Budnik's 1971 photograph of de Kooning in his studio, for example, shows a bottle of Saff-o-life safflower oil on the artist's work table (fig. 44).[98] While Thomas Hess was the first of de Kooning's contemporaries to report that the artist added this oil to his paints, visitors to the artist's studio remember that he boasted that he had found a cheap salad oil that he could use in lieu of more expensive artists' oils (T. Hess 1967, 32).[99] In a 1978 interview, de Kooning said that he used safflower oil because it "stays wet a long time, it doesn't dry like linseed oil, I can work longer" (Rodgers 1978, 20).

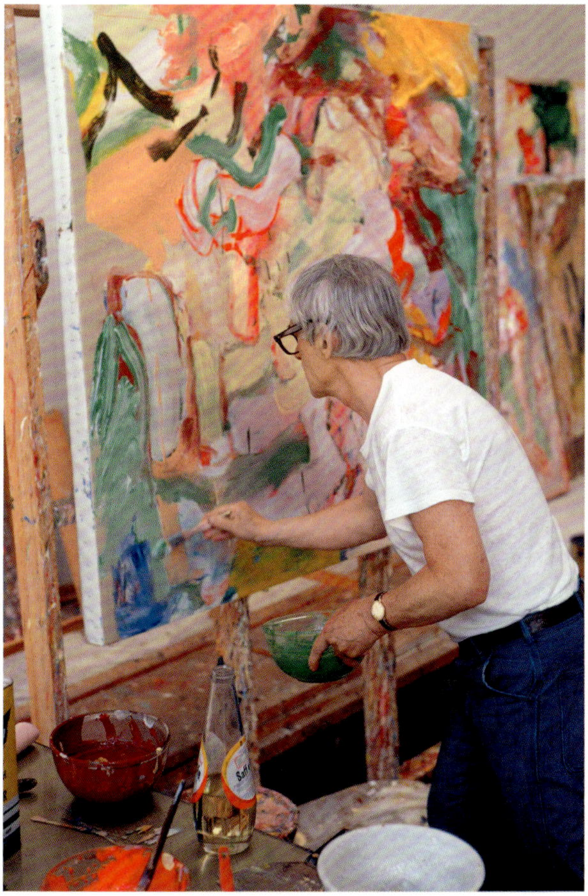

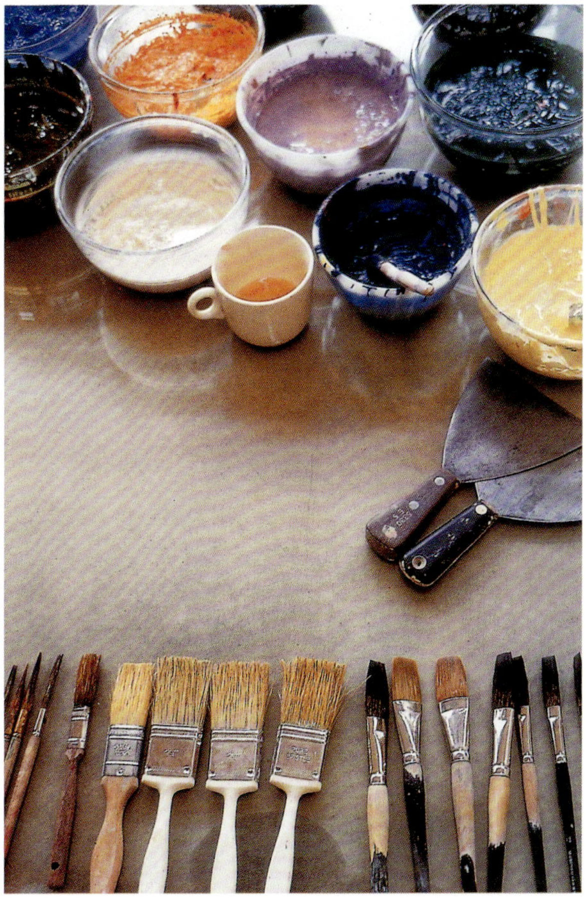

FIGURE 44
De Kooning in his studio, Springs, East Hampton, 1971; a bottle of Saff-o-life safflower cooking oil is visible on the artist's worktable.

Artwork © 2010 The Willem de Kooning Foundation /Artists Rights Society, New York. Photo: © Dan Budnik

FIGURE 45
De Kooning's worktable, Springs, East Hampton, Fall 1971.

Photo: © Dan Budnik

According to John McMahon, de Kooning prepared his paint mixtures from Bellini artists' tube colors following specific recipes that he had developed (McMahon 1993).[100] Once the artist thoroughly combined the desired number of tube paints on a glass palette, he scooped the blended paints into a bowl and then added safflower cooking oil, water, and kerosene or some other solvent, whipping the ingredients with a brush to a fluffy consistency (fig. 45). He did not add the safflower oil or the water according to a set formula but rather "by feel," until the paint achieved the desired consistency. Once prepared, he covered the paint mixtures with plastic wrap. They remained workable for several weeks and sometimes up to a month. Before turning to safflower oil, de Kooning had experimented with a number of different vegetable oils. He eventually consulted a chemist, who recommended safflower oil and assured him that he could safely add it to his paints. He purchased the cooking oil from a nearby grocery store, and when their supplies ran out, he had it specially ordered (McMahon 1993). Later, when General Foods discontinued its production of Saff-o-life, de Kooning purchased the "right" safflower oil from art-supply manufacturer Leonard Bocour in New York (Rodgers 1978, 20).[101]

Medium analysis corroborates these reports. The P/S ratios from eight paint samples taken from *Woman, Sag Harbor* and six taken from *Woman* (1965) all fall between 2.3 and 3.8, most of which correlate with two-year-old, naturally-aged standards run of pigmented and unpigmented safflower oil. Likewise, the P/S, A/G, and P/G molar ratios of two paint samples taken from *The Visit* (1966–67) and three samples from *Two Figures in the Landscape* (1967) correlate with safflower oil. But these values are also consistent with those of mock-up paint samples, prepared by the author, of artists' tube paints with a linseed oil binder combined with 50–100%, by weight, unpigmented safflower oil.[102] Thus, based purely on peak area and molar ratios, these values can be interpreted as pure safflower oil or mixtures of large amounts of safflower oil with linseed oil. Likewise, they may also be interpreted as walnut oil or soybean oil or as mixtures of poppy oil and linseed or sunflower oils. The higher P/S values of the paintings of the 1960s could also reflect mixtures of poppy oil and safflower oil. Thus, neither fatty acid nor glycerol analysis can conclusively differentiate mixtures of different drying oils. The binding medium of the paintings of the 1960s, based solely on gas chromatographic analysis, can be identified only cautiously as a mixture of linseed oil and safflower oil. In the end, however, the analytical results in combination with photographic and anecdotal evidence are sufficient to conclude with some assurance that de Kooning employed an unorthodox medium of tube paints mixed with unpigmented safflower oil.[103]

Cholesterol, a component of egg used in mayonnaise and in egg tempera medium, was not detected in any of the methylated or silylated samples, confirming de Kooning's assertion that he never added mayonnaise to his paints.[104] Glycerol analysis also supports de Kooning's statements. Egg yolk can be readily detected in oil paint because its P/G molar ratio, 1.2, is much higher than that of the typical drying oils (Schilling and Khanjian 1996, 224). The notion, once widely accepted, that de Kooning's unusual medium included mayonnaise may have derived from vivid descriptions of the artist's temporary emulsions of oil and water beaten until they achieved a "fresh, creamy nature" with the "consistency of heavy sauces" or "a good mayonnaise" (T. Hess 1975, 73; 1978, 236; 1972c; Tomkins 1987, 66).

De Kooning during the 1960s seems to have been experimenting with new ways to apply abundant amounts of paint, and his unorthodox paint formula was central to the distinctive qualities of his paintings of the decade. It allowed him to rework a painting over a far longer time, and it imbued his brushwork with an expressive force that surpassed that of the 1940s. The repeated reworking was a long-established practice. But the use of a slower drying paint allowed him more easily to scrape down all or part of a picture's surface, leaving no more than traces of paint and charcoal lines. The door paintings retain only faint residues of former images, rather than the substantial build-up from earlier paint layers evident in many of the works of the 1940s and 1950s. The surfaces of these paintings give little clue as to the actual length of time it took to execute them; the final picture records only the last of many stages of painting, scraping, and repainting. In an undated letter to Joe and Olga Hirshhorn, de Kooning described his working method:

> I am working for weeks and weeks on end on a large picture . . . and have to keep the paint wet . . . so that I can change it over and over,—I mean, do the same thing over . . . and over, and that [way] it will look fluid than [*sic*] . . . and fresh. —as if it was really a small picture (Zilczer 1993b, 167).[105]

John McMahon (1993) confirmed de Kooning's extravagant use of paint, in which he regularly applied and removed the equivalent of many tubes of paint every day. Thus, the final painting retains an appearance of immediacy, but the artist worked long and hard to achieve it.

What de Kooning's mixture implies in terms of prolonged working time is difficult to determine. Drying times of oil paints can vary greatly, depending on the type of paint and the environment in which the paint is drying. A test sample prepared by the

FIGURE 46 (below left)
Detail from right of center of *Woman, Sag Harbor*, showing the artist's use of quick strokes to apply his paints and his method of wet-into-wet paint handling.

FIGURE 47 (below right)
Detail from figure's lower torso of *Woman* (1965), showing crater-like formations in the paint surfaces that resulted when the artist whipped water and additional oil into his paint.

author using a mixture of equal parts linseed oil binder and safflower oil, admixed with 25% by weight water, did not begin to form a skin until four days after it was applied. In general, de Kooning appears to have been able to blend the colors on the painting surface freely for at least three days, and for at least a week he could still scrape down the accumulated layers.[106]

The greater liquidity and slipperiness of de Kooning's binding medium and his use of smooth, nonabsorbent supports also allowed him to move his brush across the face of the painting at a quicker pace. Despite the reworking evident in both *Woman, Sag Harbor* and *Woman* (1965), which took place over extended periods of time, much of each image as we see it now was probably completed in a few concentrated sessions of work. While a quick brushstroke is a characteristic of the earlier paintings, one who views the door paintings senses the increased speed with which this more fluid paint was applied. Everywhere the eye follows the rapid movement of the brush where it hit the panel, where it skidded, twisted, and abruptly turned when it was lifted (fig. 46). Such immediacy conjures up the very act of painting. De Kooning's friend Joan Ward remembers experimenting with de Kooning's medium; she found it so slippery that she "could not keep the brush on the canvas" (Shiff 1994, 65, n. 20).

Another important effect of de Kooning's slower drying binding medium was to facilitate more complex and varied brushwork than he had used at any time earlier in his career. By the 1960s, the heavy impasto of the paint in these works was due to the properties and consistencies of the paint medium itself, rather than to the addition of gritty materials into the paints. In creating *Woman, Sag Harbor* and *Woman* (1965),

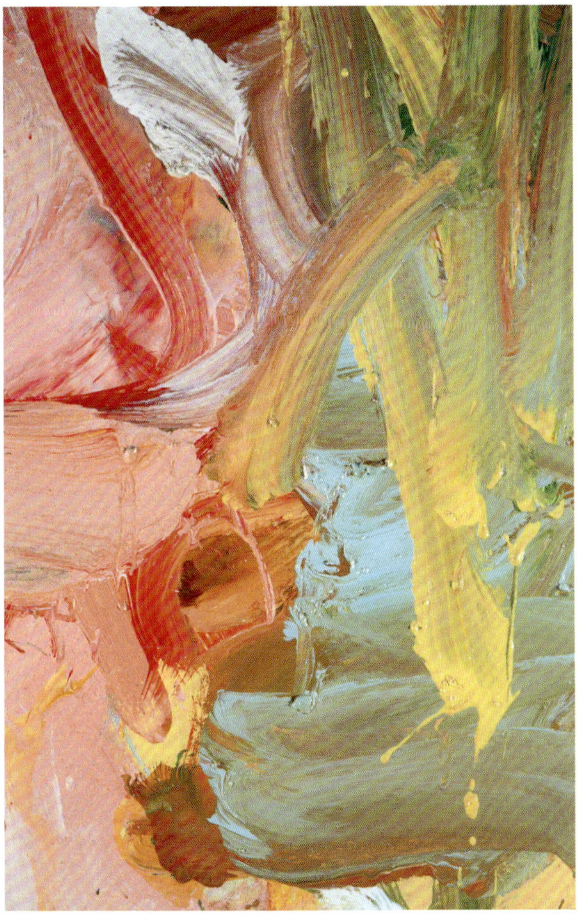

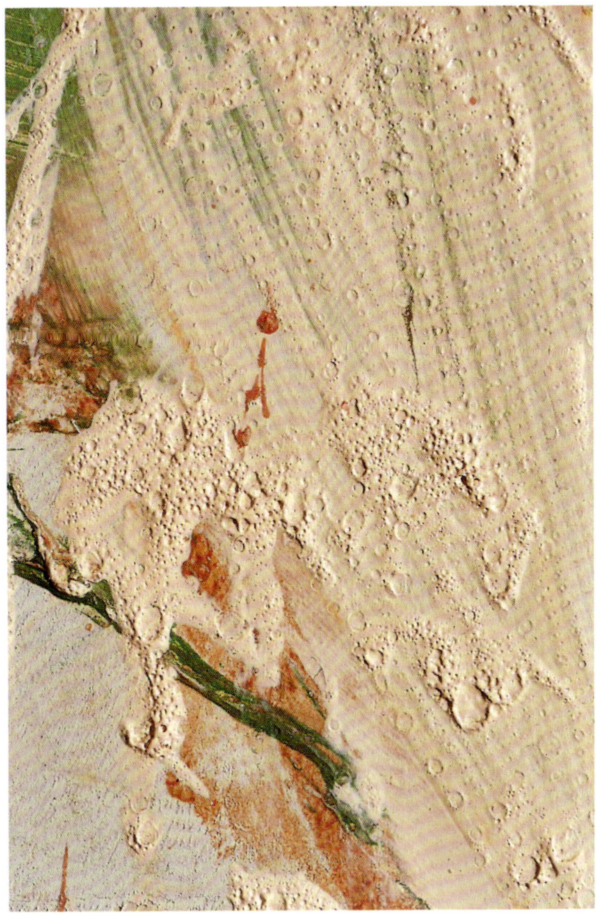

de Kooning exploited the full range of textures and color effects that his medium was capable of achieving, using a wide variety of techniques to apply and rework his paints. The oil-rich paint was worked with a true wet-into-wet technique, with later brush-strokes dragging and blending with earlier ones to produce complex textures and vivid color mixtures on the picture surface. In *Woman, Sag Harbor*, for example, de Kooning outlined the figure's torso with a brush heavily loaded with a fluid yellow paint, which he brushed through still-wet areas of pink, red, green, and blue. As he worked, he dragged in the adjacent colors, blending them on the surface.

In addition to complex paint textures created with the brush, the surfaces of *Woman, Sag Harbor* and *Woman* (1965) show peculiar circular depressions and wrinkles that formed upon drying, a common feature of many of de Kooning's paintings of the 1960s and 1970s. They seem to reflect the artist's response to the real world: the watery environment of East Hampton and the sensuality of human flesh. Circular, crater-like formations in the paint surfaces of many pictures of this period were certainly formed as water and air bubbles, formed from the temporary emulsion of water and oil, popped or collapsed once the paint was applied (fig. 47). According to Peter Rudolph, a painting student who befriended de Kooning in 1966, the artist was then mixing his tube paints with a particularly liquid medium made up of equal parts safflower oil, solvent, and water.[107] In 1964, the year both *Sphinx* and *Woman, Sag Harbor* were painted, de Kooning acknowledged that he was searching for a way to convey the very substance of water and that even the most abstract images of this period are related in some way to the watery environment of Long Island:

> Now I go on my bicycle down to the beach and search for a new image of the landscape. And I love the puddles. When I see a puddle, I stare into it. Later, I don't paint a puddle, but the image it calls up within me. All the images inside are from nature anyway (Dickerson 1964a, 16).[108]

Wrinkles are a common occurrence in the paint surfaces of works of the 1960s and they reached an extreme later in the decade in paintings such as *Two Figures in a Landscape* (1967; figs. 48 and 49). They are often discussed by art historians as a kind of "shrinkage" associated with de Kooning's use of water. In fact, these unconventional formations likely resulted from the artist's addition of excess oil to his paints and his heavy build-up of paint, as similar effects can be achieved readily with linseed oil alone. It is an old and reliable rule that wrinkling or shriveling in paint occurs as the film sets if a certain maximum film thickness is exceeded, especially in paints with a high percentage of binder and with insufficient pigment to reinforce the film. As the upper portion of the oil film oxidizes, it forms a kind of skin that expands more than the lower layer. Wrinkling is most extreme when the ground is a nonabsorbent oil layer and when water is present to inhibit oxidative drying of the lower paint layers. This disproportionately increases the volume of the surface until it assumes a kind of folded corrugation (M. Hess 1965, 83).

Although the pits and wrinkles on the surfaces of the paintings of this period may be seen as resulting from failed or defective paint, they were likely intended, and de Kooning developed his own technical procedures to achieve them. As was the case in paintings from the 1940s and 1950s, the artist purposefully disregarded established painting practices in order to achieve desired visual effects and expand his descriptive ability. Just as the artist's oil-rich medium infused *Woman* (1964) with a frank sexuality, the figures in *Woman, Sag Harbor* and *Woman* (1965) evoke human flesh itself, overtly conveying sensuality through the puckered paint film and the wet-looking medium-

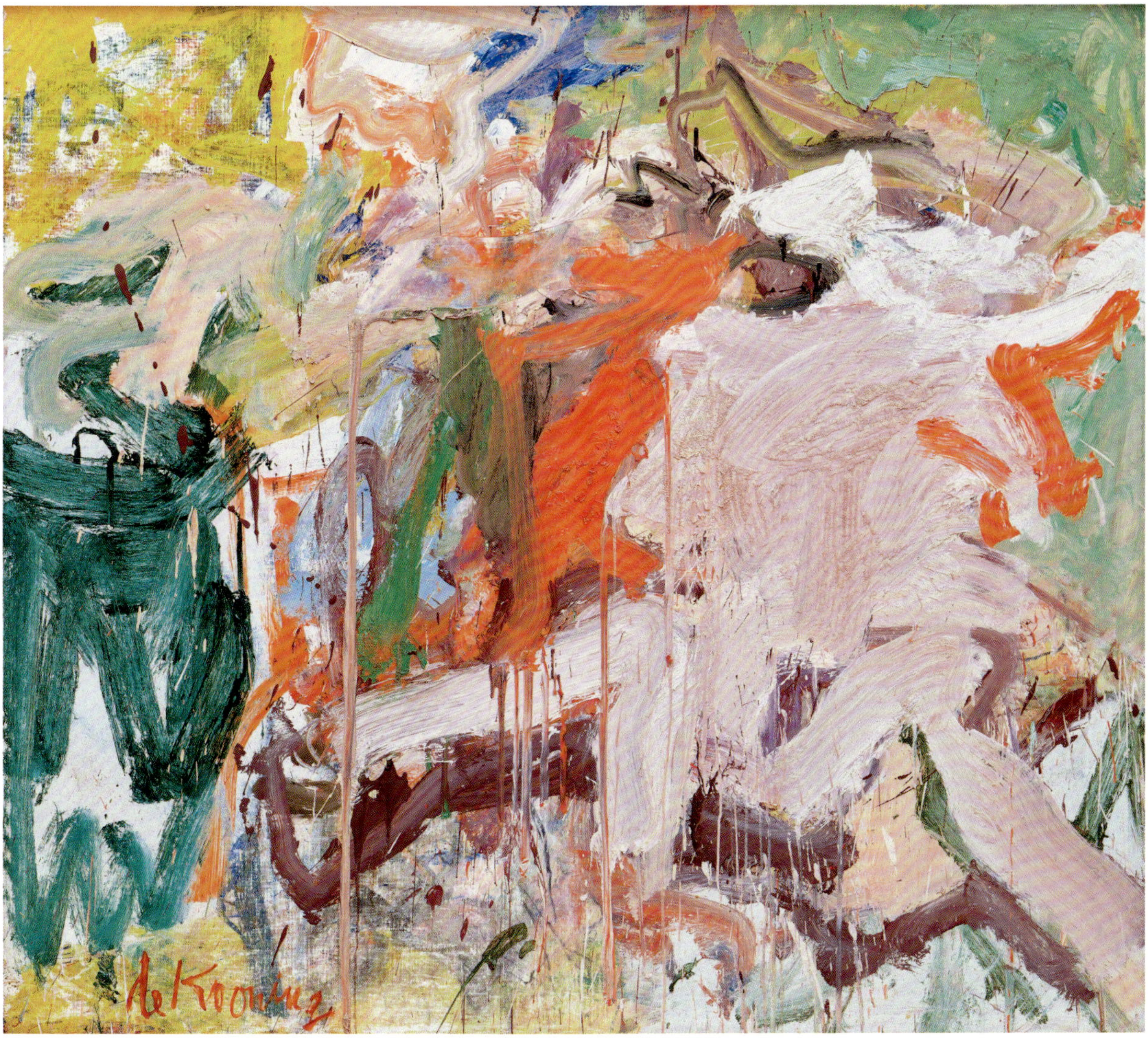

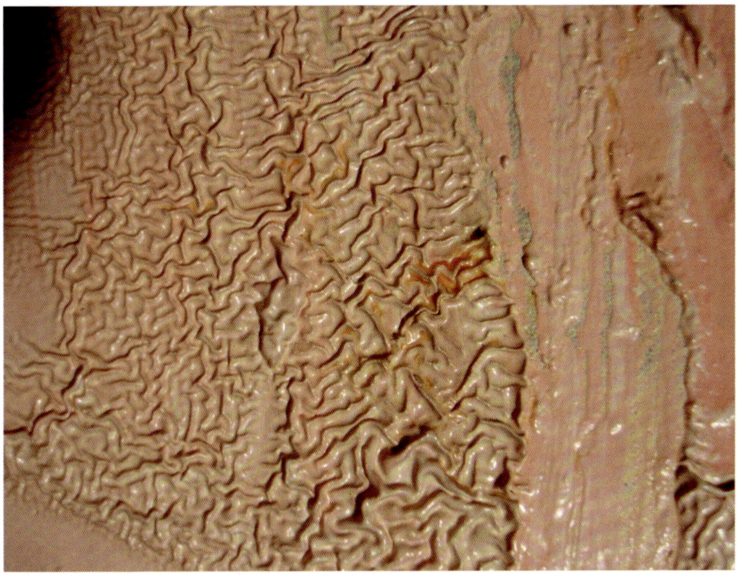

FIGURE 48
Two Figures in a Landscape, 1967. Oil on canvas, 177.8 × 203.2 cm (70 × 80 in.). Stedelijk Museum, Amsterdam.

© 2010 The Willem de Kooning Foundation / Artists Rights Society, New York

FIGURE 49
Detail from center left of *Two Figures in a Landscape*, showing drying wrinkles in the paint.

rich paints. The dimpled, wrinkled, and glistening surfaces of these paintings are associated almost exclusively with the pink paints, areas that can be equated with human flesh. Thus, these works simultaneously recall the contemporary nude and the tradition of fleshy sensuality in Western painting as de Kooning had defined it in 1951: "Flesh is the stuff people are made of... the reason why oil painting was invented" (de Kooning 1951a, 86).[109] In the paintings of the 1960s, the artist pursued this fleshiness by means of subject choice, materials, and methods alike.

Woman, Sag Harbor and *Woman* (1965) exemplify de Kooning's virtuosity in handling paint. Although elements of traditional practice remain, such as the use of preparatory drawings and carefully prepared grounds, overall these paintings reveal the innovation of de Kooning's approach. His technique is direct, loose, and extravagant and appears deceptively simple, yet the calculation of color and texture is complex and exact. By manipulating his colors and the thickness and textures of his paints, de Kooning truly managed to become, in Dore Ashton's apt phrase, "a painting wizard" (Ashton 1968, 39; Lake and Zilczer 1993, 179).

Once de Kooning moved to Long Island, he altered his technical procedures to accommodate his new environment. The supports of his paintings of the 1960s are slick papers and primed canvases or panels, sanded to a smooth finish and often white in color. His colored pigments are admixed with high percentages of white that accentuate the luminous quality of the paints. Likewise, de Kooning experimented with different binding media. He used house paints infrequently and then only during the early years of the decade. Rather, he began adding poppy and then safflower oil, mixed with water, to his artist's paints. This unusual medium increased the drying time and fluidity of his paints and facilitated a quicker brushstroke. The artist's modifications to his paints allowed him to achieve the daringly unorthodox paint surfaces that have often been compared to the flesh of sunbathers, the water, and the North Atlantic coastal light. De Kooning's modifications of his oil paints were central to the extraordinary character of the paintings of this period.

De Kooning in Context

When the paintings generated by the artists now known as Abstract Expressionists first appeared in New York City galleries, they shocked their viewers, so unlike anything in the history of art did they appear. Although scholars since have recognized the diverse roots of the movement, historians are still divided as to its legitimacy and meaning. As Stephen Polcari noted, from the beginning Abstract Expressionism has been interpreted differently depending on the culture and intellectual climate of the generation of historians and critics examining it (Polcari 1991, xvii). Technical examination and scientific analysis does not offer a way out of this knotted history of evaluation and interpretation. Rather, it offers an additional means of gaining insight into the artists' motivations. Preliminary evidence from the technical study of the works by de Kooning has established that the artist did not select his materials randomly or solely because they were cheap but rather made conscious artistic choices. Likewise, his paintings were not made up exclusively of automatic or arbitrary gestures. Rather, he knew his materials well and consciously deliberated on his methods of applying them.

Technical study of de Kooning's paintings from the 1940s and 1950s compared with those of the 1960s reveals not an abrupt repudiation of his past materials and methods but subtle, thoughtful modifications through which he shifted the emphasis of his paintings to reflect his new surroundings. He never lost his respect for the craft that went into the production of his work, and throughout his career he drew on his extensive academic training and his work as an applied artist. Sometimes he adroitly used methods and materials in ways that have been practiced for centuries. At other times, he went to considerable lengths to challenge his skills, intentionally misusing his materials and disregarding traditional procedures of painting as a means of discovering a fresh visual statement. Late in his career, he remarked, "The accumulation of these things you learn become very important" (Berman 1982, 73).

Even as he experimented, he continued to draw upon the traditional practices of the old master painters. He applied his paints with brushes to paper, to solid panels, and to canvas supports, which were often prepared with layers of sizing and priming and placed in a vertical position. Additionally, drawing was always central to de Kooning's method of working. Frequently, he laid out his composition with an initial drawing, to which he referred throughout the painting process, working directly from a large repertoire of sketches. Since at least the mid-1940s and probably earlier, he painted in fits and starts, through a process of continual reworking. Finally, de Kooning consistently worked with oil-based paints throughout his career. Unlike many of his fellow Abstract Expressionists, he never experimented with the quick-drying synthetic resin paints that became available around 1950. Rather, he carefully chose and modified his oil-based paint system—whether artist's, house, or sign paint—to suit his subject matter.

Due to these "traditional" practices and because he did not consider modern art as a break from the past, de Kooning has been described as the most "conservative" of the Abstract Expressionists (Polcari 1991).[110] He has also been credited with being the artist who did the most to extend the painterly tradition in art (Ashton 1965). Critics and artists alike have recognized de Kooning's prodigious knowledge of painting materials. As critic James Fitzsimmons (1953, 7) observed, "There is probably no one around who knows more about the techniques of painting than de Kooning." Joos Sanders, a friend and a fellow Dutch-born artist, averred that de Kooning knew more than any of his peers about traditional commercial art techniques and was the most inventive in incorporating such methods into his paintings (Stevens and Swan 2004, 186).

This study opened with the premise that artists achieve radical changes in style through deliberate changes in their artistic procedures. Technical examination and analysis of de Kooning's work from the 1940s through the 1960s has identified a number of points in his painting practices where he consciously modified more conventional painting techniques. Thus, the decisive changes in style in paintings from these three decades represent a deliberate turn in a new direction and the materials that he chose to use are inextricably linked with his pictorial evolution.

De Kooning, along with Jackson Pollock and Franz Kline, represented the so-called gestural or "Action Painters" of the Abstract Expressionist movement. While de Kooning and Jackson Pollock were the best-known Abstract Expressionists during the early years of the movement, they were not working in isolation. De Kooning's work should be viewed in the context of widespread changes in painting practice and concurrent changes in technique that he may have shared with his contemporaries. When we have similar information on the technical practices of the other Abstract Expressionist artists will we be able to quantify the degree to which the individual painting styles by these diverse artists were dictated by their different painting procedures. Only then will we be able to place the achievements of each artist within the context of the group and more accurately assess the innovations of the Abstract Expressionist movement as a whole.

4 Analysis of de Kooning's Media in Paintings from the 1960s and 1970s

Susan F. Lake and Michael Schilling

The goal of the technical analyses described here is to determine why some of de Kooning's paint films have dried normally while others have not, and to determine whether there is a relationship between the composition of the paints and their stickiness. Medium analysis was done on sixteen paintings executed between 1960 and 1977.

The four works from the 1970s were included in this study because many of de Kooning's paintings executed prior to about 1978 have condition issues similar to those executed during the 1960s. Table 3 of Appendix 1 lists these four paintings and gives the results of medium analysis. Table 2 of the same Appendix cites the twelve works examined and analyzed from the 1960s. The pigments of six of the paintings executed during the 1960s were also identified, although pigment analysis was not possible for all of the paintings of the 1960s or for any of those from the 1970s, due to the small sample sizes. This project represents a collaborative effort: medium analysis was carried out in a number of laboratories using various analytical procedures. Details of the methodologies are given in Appendix 2.

Because the six paintings in which both pigments and the binding media were identified have neither been cleaned with solvents nor subjected to lining, changes in the composition of the oil media must be due, for the most part, to natural aging processes. This allows for meaningful comparison without having to factor in the effects of conservation treatments.

Many of the paints in de Kooning's works that were executed up to about 1964 have remained soft or gooey for more than forty years. Soft and malleable paints are, in fact, a common characteristic of de Kooning's paintings well into the 1970s. In *Woman, Sag Harbor* (1964) and *Woman* (1965), the paints that have remained soft generally are the full-strength cadmium colors or those with cadmium pigments combined with synthetic organic dyes. By contrast, the paints with significant quantities of white pigment, especially those containing zinc oxide, have formed hard films.

For the purpose of this study, hard samples are defined as those that cannot be deformed by the point of a needle, while soft samples are those that have a uniform consistency throughout the paint film, but which can be permanently deformed by a needle. Gooey paint is paint that has formed a surface skin, but retains a near liquid interior. Gooey paint can be deformed with a needle but regains its original shape within about 30 minutes. Moreover, gooey paint cannot be removed from the painting as a discrete shape, but forms taffy-like strings once the paint film skin is broken. If crater-like voids were present in a paint sample, this was taken as evidence that de Kooning added water to the paint.

Chemistry of Drying Oils

Although the drying chemistry of oils in painting has been discussed in great detail in other publications, it may be useful to recap a few essential concepts. Fats and oils are composed of triglycerides, which are esters of one molecule of glycerol with three molecules of fatty acids. The compositions of some common oil paint media are listed in Table 4 (Appendix 1).

In chromatographic analysis of oil paints, saturated fatty acids (palmitic acid and stearic acid) are relatively inert marker compounds, whose relative abundance, abbreviated as P/S, is useful for identifying traditional drying oils (Mills 1966; Mills and White 1994). P/S ratios between 1 and 2 are indicative of linseed oil, whereas ratios near 5 are characteristic for poppy oil. Paints with P/S ratios near 3 may contain walnut oil, safflower oil, or a mixture of poppy oil and linseed oil (Schilling and Khanjian 1996).

Unsaturated fatty acids enable oils to combine with oxygen to form hard, dry, highly cross-linked films (in a process called autoxidation). Linolenic acid reacts with oxygen about three times faster than linoleic acid, whereas linoleic acid reacts ten to twenty times faster than oleic acid. Accordingly, the overall drying rate of oils is affected by the relative amounts of unsaturated fatty acids. For example, linseed oil contains 48–60% linolenic acid and dries quite rapidly, whereas the slower drying safflower and poppy oils contain 14–19% linoleic acid with trace amounts of linolenic acid.

Some of the reactions that occur simultaneously with polymerization lead instead to bond breaking, with the formation of lower molecular weight compounds such as dicarboxylic fatty acids. Azelaic acid (C_9) is by far the most abundant dicarboxylic fatty acid in dried oil paints, but suberic acid (C_8) and sebacic acid (C_{10}) are also significant. Depending on the sample preparation procedure, dicarboxylic fatty acids from C_2 to C_7 can also be detected by gas chromatography/mass spectrometry (GC/MS) in lower concentrations.

Another important reaction that leads to depolymerization, the hydrolysis of fatty acid glycerol esters to form free fatty acids, has been the subject of intense scientific investigation (Van den Berg 2002). Free fatty acids present in the original paint film, plus those formed during hydrolysis, subsequently may combine with basic pigments that contain lead, zinc, or cobalt to form metal soaps. Exploration of this process, which is a major factor in the drying and overall stabilization of oil paints, led to the development of an ionomeric model for aged oil paints (Boon et al. 2002).

Besides soap formation (saponification), other physical and chemical processes may deplete free fatty acids from the paint film. These mechanisms include migration of free fatty acids into neighboring paint layers or the painting support, and the formation of crystalline bloom on the painted surface (Koller and Burmester 1990; Moffat and Miller 1996; S. Williams 1989). Furthermore, fatty acids are sufficiently volatile at room temperature that they may evaporate from paint layers, which is one cause of so-called ghost images on glazed easel paintings (S. Williams 1989; Michalski 1990; Schilling, Carson, and Khanjian 1997).

Each drying oil has a well-defined and measurable ratio of palmitate to glycerol (P/G) and, moreover, the P/G ratio of an oil paint is at its maximum value when the paint is fresh. For example, Table 4 in Appendix 1 shows the mean P/G molar ratio is 0.03 for castor oil, 0.18 for linseed oil, 0.21 for safflower oil, and 0.33 for poppy oil. As oil paints age, free fatty acids are liberated by hydrolysis of the glycerol esters and are subsequently removed by the chemical and physical processes described earlier. Depending on the extent of glyceride hydrolysis and the amount of free fatty acid depletion, decreases in the P/G ratio of an oil paint should be observable over time.

Paint Mock-ups

It was decided to examine de Kooning's unusual paint media to determine whether the safflower oil alone, the water, the pigments, or any combination of these could explain the different drying phenomena of the artist's binding medium. Paint mock-ups were prepared according to John McMahon's descriptions of de Kooning's mixtures. The mock-ups (MU) for this study were prepared with a set weight of an artist's tube paint (Winsor and Newton cadmium red light, ground in alkali-refined linseed oil) to which safflower cooking oil (Hain Food Group, Inc., Garden City, N.Y.) was added in increments of 10 weight % to cover a range of 10 to 100%. A duplicate set of samples was prepared with the addition of 25 weight % of water. The mock-ups were applied to glass plates and aged under ambient conditions of temperature and humidity for three years prior to analysis.

Solvent Extraction Studies

Samples from *Woman, Sag Harbor*, from *Woman* (1965), and from the paint mock-ups were analyzed for free fatty acids (as trimethylsilyl esters) and glycerides (as methyl esters) using GC/MS following a stepwise extraction and derivatization procedure (Koller and Burmester 1990) (Appendix 2E). A hard white paint sample was taken from *Woman, Sag Harbor*, and a hard pale pink sample was removed from *Woman* (1965). A soft orange paint sample was taken from the proper left shoulder of the figure in *Woman* (1965). A maroon sample from the genitalia of *Woman, Sag Harbor* was so gooey that it was removed from the painting in a series of strings wrapped around a tungsten needle, then transferred to a glass slide. The composition of the pigments in these paint samples is listed in Table 2 of Appendix 1.

The mock-ups were analyzed for comparison with the samples from the paintings. These included a sample of artist's tube paint with a linseed oil binder taken directly from the tube (MU1), a sample of the same artist's tube paint mixed with an equal amount of safflower cooking oil (MU8), and a sample with equal amounts of the artist's tube paint and safflower oil admixed with 25 weight % of water (MU22).

In the GC/MS results for the free fatty acid and total glyceride tests, no qualitative differences were observed between any of the paints—each chromatogram showed peaks for dicarboxylic fatty acids, palmitic acid, stearic acid, oleic acid, and linoleic acid. The total amount of medium extracted from the two de Kooning paints and from the mock-ups to which safflower oil had been added was relatively high, as would be expected in a film to which a good quantity of oil was added. In each de Kooning paint sample examined, the proportion of medium to pigment was between 40 and 60%.

The results of the benzene/hexane extraction of the mock-ups revealed that the paint films to which safflower oil was added—in particular the samples admixed with safflower oil and water—have significantly higher amounts of free fatty acids relative to the pure linseed-oil paint sample. The resulting chromatograms of the three mock-up paint samples show the marked effect that safflower oil, in combination with the water, had on the formation of free fatty acids (fig. 50). This same general trend can be seen in the chromatograms of the silylated samples of the gooey maroon paint relative to the hard white paint sample from *Woman, Sag Harbor* (fig. 51).

Table 5 in Appendix 1 presents peak area ratio results for the paint mock-ups and the de Kooning samples. All of the paint samples mixed with safflower oil show elevated proportions of free fatty acids (FFA) relative to the unmixed tube color.

FIGURE 50

Gas-chromatograms of free fatty acids extracted and silylated. Top: cadmium orange artist's paint straight from the tube with linseed oil binder (MU1). Middle: 1:1 mixture by weight of artist's tube paint with safflower cooking oil (MU8). Bottom: 1:1 mixture described above with the addition of 25%, by weight, water (MU22).

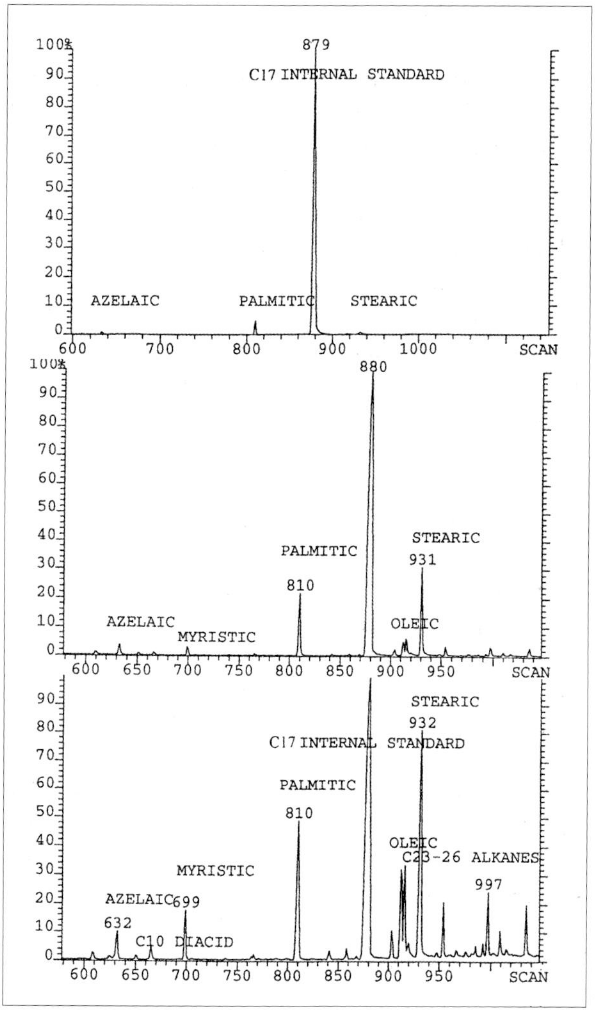

For example, the free fatty acid content of the paint mixed with an equal amount of safflower oil (MU8) was nearly 6 times higher than that of the unmixed tube color (MU1), and the sample with both safflower oil and water (MU22) added had nearly 18 times the free fatty acid content of the unmixed tube color. Significantly, the hard samples from both door paintings and the soft orange sample from *Woman* (1965) had peak area ratios similar to the mock-up paint sample with safflower oil added and the mock-up with both safflower oil and water added. By contrast, the ratios of the gooey maroon sample from *Woman, Sag Harbor* indicate that this paint has almost 6 times the free fatty acids relative to that in the hard white sample from the same painting and almost 35 times that of the unmixed tube color.

Quantification of the peak area ratios of the silylated fractions and comparison of these with the glyceride fractions from the same samples give some indication of the degree of hydrolysis of each oil paint sample and, conversely, the extent of cross-linking of the oil network in the paint films (Appendix 1, Table 6). As would be expected, samples with the highest proportion of free fatty acids also have the lowest amount of glycerides remaining in the paint matrix. For example, the free fatty acid content (FFA) of the unmixed tube color is roughly 10% of the total fatty acid content in the paint film, and the free fatty acid content of the mock-up with an equal amount of safflower oil added is about 13% of the total fatty acid composition of the film, not an

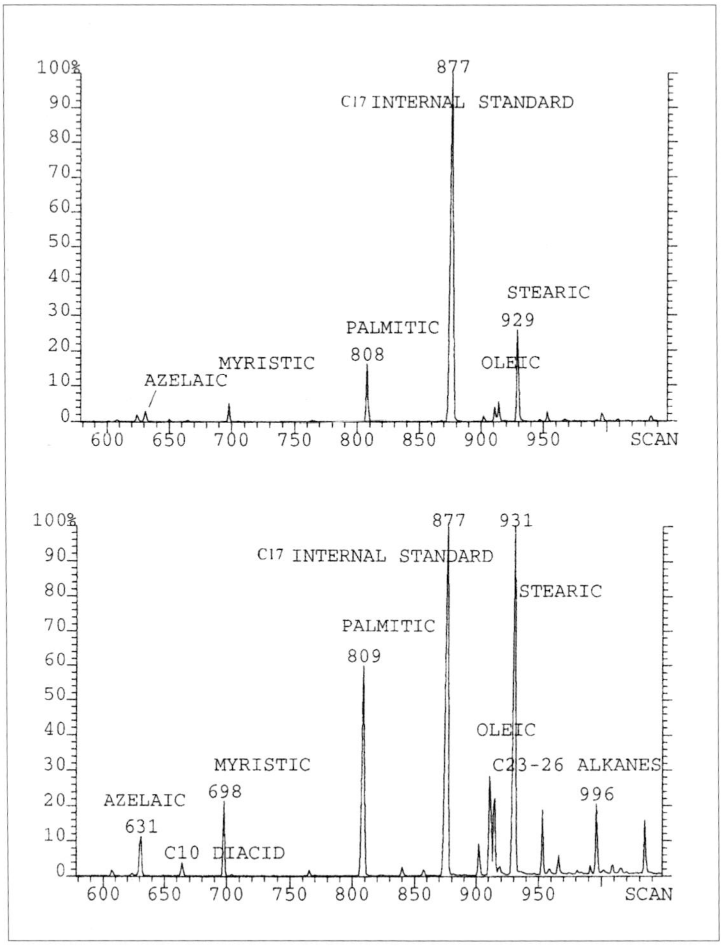

FIGURE 51
Gas-chromatograms of free fatty acids extracted and silylated from two paint samples taken from *Woman, Sag Harbor*. Top: a hard white sample from lower left corner. Bottom: a very gooey maroon sample from figure's genitals.

appreciable difference. By contrast, the free fatty acids extracted from the sample with both safflower oil and water added to the paint is significantly higher, measuring about 34% of the total fatty acid composition of the oil film. These percentages are remarkably similar to those of the hard white sample from *Woman, Sag Harbor* and the pink sample from *Woman* (1965), with fatty acid compositions of 14% and 22%, respectively. The softer orange sample from *Woman* (1965) has a free fatty acid component that is 32% of the fatty acids in the oil film. Only the gooey maroon paint from *Woman, Sag Harbor* far exceeded the values of the mock-ups, with nearly 83% of its total fatty acid content in the form of free fatty acids.

In summary, analyses of the mock-up paint samples and samples from de Kooning's door paintings strongly suggest that added water and safflower oil may have retarded the drying process. Clearly, the addition of safflower oil alone to the tube paint leads to the formation of a slightly higher percentage of free fatty acids, but it is primarily the safflower oil in combination with water that contributed to increased hydrolysis of the glycerides, which leads to the formation of free fatty acids. Moreover, slow-drying oils such as safflower and poppy oils are known to be more susceptible to hydrolysis than faster drying oils such as linseed oil. Because many of these free fatty acids in a pure state are liquid at room temperature, they remain mobile in paint films, and their presence and quantity have a decisive influence on the behavior of the paint. Rather than forming bonds that lead to a solid paint film with a highly cross-linked polymer network, hydrolysis is the more common mechanism, leading instead to the formation

FIGURE 52
Gas-chromatography/flame ionization detection (GC/FID) results for total acid and glycerol analysis, soft white paint from *Amityville* (1971).

FIGURE 53
Gas chromatography/mass spectrometry (GC/MS) results for free fatty acid analysis, brittle blue paint from *Spike's Folly II* (1960).

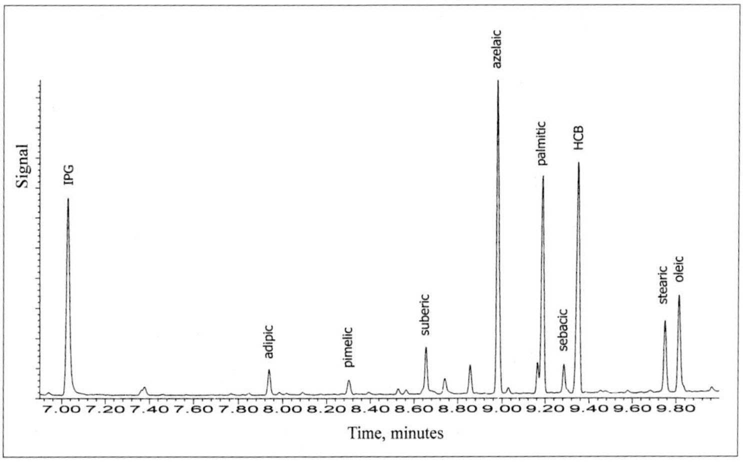

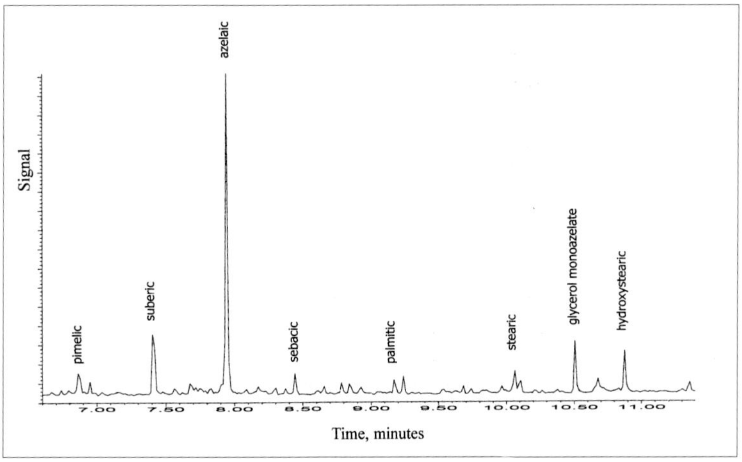

of a lower molecular-weight network that remains sticky even years after the painting was executed.

Glycerol and Fatty Acid Measurements

Additional GC/MS tests were carried out to better understand the compositional changes that have occurred in de Kooning's paintings from the 1960s and 1970s. Twenty-three de Kooning paint samples were prepared using a procedure, described in Appendix 2F, that permits quantitative analysis of glycerol and total fatty acids—that is, the sum of fatty acids present as free fatty acids, bound as glycerides, plus those present as pigment soaps (Schilling and Khanjian 1996; Mason et al. 1964). Figure 52 shows a typical chromatogram from the GC analysis of fatty acids and glycerol. The analytical results of this procedure were expressed in terms of molar ratios. One advantage of this GC procedure is that it enables one to estimate the oil content of the paint sample. Because the amount of glycerol in most drying oils is 10% by weight, the percentage of oil in a paint sample is approximately ten times the percentage of glycerol.

In a second GC/MS procedure (Appendix 2G), free fatty acids and soaps in a duplicate sample set were analyzed as trimethylsilyl esters (Schilling, Carson, and Khanjian 1997; Schilling 2005). The extent of hydrolysis in an oil-paint sample could be estimated by measuring the amount of free azelaic acid and soaps relative to the total amount of azelaic acid detected (free plus bound plus soaps). Although any fatty acid in the sample could be chosen for this estimate, azelaic acid was selected because it was present in the highest concentration in every sample. This process is similar to the one used by Van den Berg et al. (2002) to estimate the degree of hydrolysis, but it requires two GC/MS tests instead of one.

To complete the testing, four additional samples were prepared following a traditional procedure (Appendix 2C) that employs saponification and subsequent methylation using diazomethane (Mills 1966; Mills and White 1994).

Table 6 (Appendix 1) lists the main analytical findings for the set of de Kooning paint samples, which include molar ratio data (P/S, P/G), oil content, and percent hydrolysis (% FFA). It should be noted that all of the paints are assumed to contain drying oils because their A/P (azelaic to palmitic) ratios exceed 1.0. Results in the tables are arranged into three categories: samples with low P/S (2.0 and below), high P/S (3.4 and above), and intermediate P/S. Due to limited sample sizes, multiple tests of each paint sample could not be performed, thereby limiting statistical evaluation of the results.

The media in the low P/S paints were identified as linseed oil, castor oil, or mixtures thereof. Paints in the high P/S group were suspected of containing either pure poppy oil or admixtures with linseed oil. From the previous discussion, it is reasonably certain that the paints in the intermediate P/S group contain mixtures of linseed oil with safflower oil, although walnut oil cannot be ruled out entirely.

On the basis of their extremely low P/S ratios, the medium in a yellow sample from *Spike's Folly II* (1960), and a soft black sample from *Untitled V* (1977), was identified as castor oil or a binding medium containing significant amounts of castor oil. Ricinoleic acid, present in castor oil, was detected in *Untitled V* (1977), *Untitled I* (1977), . . . *Whose Name Was Writ in Water* (1975), and *Door to the River* (1960). In addition, a peak identified as 12-hydroxystearic acid was detected in the silylated extracts from low P/S samples from *Spike's Folly II*, *Untitled V*, and in a sticky yellow paint sample from *Rosy-Fingered Dawn at Louse Point* (1963). 12-hydroxystearic acid was also identified in samples with high P/S ratios from *Spike's Folly II* (blue) (as shown in fig. 53), and a soft maroon sample from . . . *Whose Name Was Writ in Water*. Silylated hydroxyl-substituted dicarboxylic fatty acids were also detected, which is consistent with the presence of hydrogenated castor oil. Also, the P/G ratios for these paints are all somewhat lower than expected, most likely because of the inherently low P/G ratio for castor oil and hydrogenated castor oil. Castor oil may be added to tube paints to increase drying times, whereas hydrogenated castor oil acts as a stabilizer.

Mean oil content was calculated from the % glycerol data for each of the three P/S categories. For the low, mid, and high P/S paints, the average oil contents are 54, 42, and 29% by weight, respectively. With few exceptions, the low P/S paints had the highest oil contents, whereas the high P/S paints contain the lowest amounts of oil. In general, the lighter colors had the lowest oil content because they contain greater amounts of zinc white and titanium white, two pigments with low to medium oil indices (Mayer 1981). In contrast, the paints with high oil content contained larger amounts of high oil index pigments such as cadmium colors, ultramarine, and alizarin.

Regarding the degree of hydrolysis in the paint samples, free fatty acids (and/or pigment soaps) were detected in all of the paints. The content of free azelaic acid (%FFA) in the low, mid, and high P/S paint groups averaged 38, 27, and 24%, respectively. As discussed earlier, evaporation of free fatty acids complicates the estimation of the degree of hydrolysis. In one study (Schilling, Carson, and Khanjian 1997), it was shown that approximately 40 years is required for palmitic acid to evaporate to half its original amount (thus, 40 years is the "half time" of evaporation), versus 80 years for azelaic acid and 120 years for stearic acid. Accordingly, in 40-year-old paintings only 75% of the free azelaic acid that was liberated early during the drying process should remain, compared to half of the free palmitic acid. One interesting observation was that ghost images formed over most of the de Kooning paint samples on the inside of the microscope slide sample holders, indicating that, even today, they continue to lose free fatty acids through evaporation. This could explain why the content of free palmitic acid was always somewhat lower than the amount of free azelaic acid in the de Kooning samples.

To summarize, the combined quantitative GC/MS glycerol and fatty acid protocols permit the quantitative assessment of oil content and degree of hydrolysis. This information was useful for assessing oil content and degree of hydrolysis of the de Kooning paint samples. The oil content and degree of hydrolysis of the low P/S paints were found to be higher than for the high P/S paints, although there appears to be no clear correlation between the age of the painting and the degree of hydrolysis. Finally, evaporation of free fatty acids has significantly altered the original composition of the oil glycerides. It is anticipated that evaporation will continue to occur, which could cause disturbing ghost images if the paintings are glazed.

Effects of Pigments

If one assumes that de Kooning added safflower oil and water to all, or at least most, of his paints for a period of many years, as has been suggested by conversations with studio assistant John McMahon (1993), the fatty acid analysis described above does not in and of itself explain why some of the artist's paints in works of the 1960s and 1970s have remained gooey while others have formed soft paints or have dried to a hard film. Causes other than the medium, therefore, were considered as a more complete explanation for the state of de Kooning's paints, namely the pigmentation of the paints.

As mentioned earlier, polarized light microscopy (PLM), elemental analysis (SEM–EDS), and x-ray diffraction (XRD) of the pigments present in several of these works (Appendix 2A, 2H, and 2J) reveal significant but variable amounts of the white pigments titanium dioxide, zinc oxide, and barium sulfate in the pastel tones and almost pure cadmium pigments in the full-strength red, oranges, and yellows. Significantly, over the range of colors throughout *Woman, Sag Harbor* and *Woman* (1965), the amount of zinc oxide relative to the amounts of titanium dioxide and barium sulfate correlates inversely with the degree of tackiness of the paints, confirming that there appears to be a relationship between the still-wet nature of the paints and the type of pigments present. Barium sulfate is inert in drying oil and titanium dioxide is at best quite inert or even has a retarding effect on the drying process. Likewise, the cadmium colors are nonreactive pigments.

Zinc oxide, by contrast, is a reactive pigment that can "consume" free fatty acids, particularly the palmitic and stearic acids, in the medium by the formation of zinc soaps. This reactivity initiates an additional film formation process that increases the hardening and toughness of the paint film (Patton 1973). Thus, the abundant presence of zinc white in the harder de Kooning paint samples analyzed may explain why the white and pastel paints have dried to a satisfactory state while the full-colored reds, oranges, and yellows with only small amounts or no zinc present remain soft or gooey. When not mixed with a basic pigment such as zinc oxide, which reacts with the free fatty acids by metal soap formation, these acids remain mobile in the paint layer.

Anecdotal evidence suggests that as time went on de Kooning became increasingly aware that his unorthodox binding medium might result in long-term instability for his paints. Gallery owner Allan Stone recalled the artist's account of a visit sometime in the late 1970s to the conservation laboratory of the Metropolitan Museum of Art. As John Brealey, chairman of the Department of Painting Conservation, was showing de Kooning the museum collection, he warned the artist that his use of semi- or nondrying oil and his addition of water to his paints were dooming his pictures to drastic and irreversible deterioration. According to Stone, de Kooning was so discouraged by Brealey's report that he stopped painting altogether for an extended period of time (Stone 1998). When he resumed, he radically changed his painting procedures, moving away from the heavily impastoed surfaces of the 1960s and 1970s. He no longer whipped his paints in bowls with the addition of safflower oil and water, but worked with paints squeezed directly onto the canvas from a tube or mixed on a palette. Once the paint was applied to the canvas, he scraped most of it off with a wide spatula, leaving behind thin, translucent veils of color.

Conservation Implications of de Kooning's Painting Technique

Despite Brealey's dire prognosis, most of the paintings of the 1960s examined for this study have maintained much of the vibrancy they must have had when they were first executed. Care must be exercised in their handling, exhibition, and storage, however, as most of these pictures have areas of soft paint. (These areas of soft paint can be identified readily with microscopic examination.) In particular, the soft surfaces of the paintings of this period are prone to picking up dust and dirt, and finding a cleaning method that does not extract large quantities of uncross-linked or lightly cross-linked substances from the medium is problematic. While these works could be glazed, many of them are too large for this to be practicable and there is reasonable probability that ghost images would form on the interior side of the glass. It is recommended, therefore, that they be displayed and stored in as dust-free an environment as possible. When these paintings travel, they must be housed in specially constructed travel frames that ensure that nothing will come in contact with their surfaces, as any material, even silicon-coated paper or Mylar, will stick to them.

Although the paints of de Kooning's paintings from the late 1940s and 1950s tend to be brittle, most of the works from this period examined by one of the authors (Lake) are in good condition. This may be due to the fact that many of them are executed on solid supports or on paper that has been solidly mounted. Only one of the works of this decade included in this study has a surface coating as a result of a conservation treatment, and the application of a varnish to these works would surely be antithetical to de Kooning's aesthetic intentions. As with the pictures of the 1960s, the cleaning of these paintings is problematic. The unfixed charcoal that the artist smeared into the paint surfaces can be abraded easily. Likewise, the quartz and glass fragments that are embedded in the paints can readily be dislodged. Moreover, the plaster of Paris that de Kooning mixed into his paints is vulnerable to aqueous cleaning agents. As a result, the works with the most vulnerable surfaces may not be cleaned, other than selectively with a soft brush and under magnification. As with the later works, glazing is an option for the smaller paintings of this period. When this is not possible, storage and exhibition of these works in a dust-free environment is advisable.

Appendix 1

Tables

TABLE 1 Analysis of de Kooning's Paints in Works from the 1940s and 1950s

Painting	Sample Site	Pigments	Binding Medium
Queen of Hearts 1943–46	ground	calcium sulfate titanium white zinc white[b]	protein-based glue
	green w/ some white, proper right arm	organic blue/green lead white titanium white	drying oil (P/S 1.3)[#]
	orange, from shoulder	cadmium red lithopone cadmium yellow calcium sulfate titanium white	linseed oil (P/S 1.7)[#]
	pink, bottom right edge	cadmium red lithopone organic red titanium white zinc white	drying oil (P/S 1.4)[#]
Special Delivery, 1946	yellow, top edge, center	cadmium yellow titanium dioxide barium sulfate[b] plaster of Paris	retail trade paint: linseed oil (P/S 1.9) + colophony[+]
	white, lower left corner	plaster of Paris	retail trade varnish: soybean or safflower oil (P/S 2.5) + large amount of colophony[+]
Valentine, 1947	white, lower left	titanium dioxide	not analyzed
	pink, from heart	titanium dioxide organic red cadmium red[b] iron oxide[b] quartz	not analyzed
	beige, left of center	titanium dioxide iron oxide[b] carbon black[b] quartz	not analyzed
	black, lower left corner	sulfur (predominantly)	not analyzed
Untitled, 1947–48	white	titanium dioxide organic red[b] cadmium red[b] cadmium yellow[b] iron oxide[b] plaster of Paris calcite glass	not analyzed

continued

TABLE 1 (CONTINUED)

Painting	Sample Site	Pigments	Binding Medium
Zurich, 1947	white, center of left edge	titanium dioxide zinc oxide organic yellow cadmium orange[b]	retail trade paint: linseed oil (P/S 1.7) + large amount of colophony[+]
	black, lower right corner	carbon black sulfur[b]	retail trade enamel: oil-ester gum (estimate) (P/S 2.8)[*]
Black Friday, 1948	white, bottom edge, near signature	not analyzed	retail trade paint: linseed oil (P/S 1.9) + colophony[+]
	black, top edge	not analyzed	retail trade enamel: oil-ester gum (estimate) (P/S 2.1)[*]
Secretary, 1948	yellow, below orange form	cadmium yellow	linseed oil (P/S 1.5) + wax[+]
	gray, near lower left corner	titanium dioxide clay carbon black[b] quartz glass	retail trade paint: oil-modified alkyd[+]
	black, right edge center	not analyzed	retail trade paint: tung or castor oil (P/S 1.0) + large amount of colophony[+]
Woman, 1948	white, center of left edge	titanium dioxide zinc oxide plaster of Paris cadmium yellow[b]	retail trade paint: linseed oil (P/S 1.9) + colophony[+]
	gray, near figure's left leg	titanium dioxide zinc oxide iron oxide[b] carbon black[b] plaster of Paris quartz	retail trade paint: linseed oil (P/S 1.8) + colophony[+]
	pink of skirt	titanium dioxide organic red brown glass	retail trade paint: linseed oil (P/S 1.6) + colophony[+]
	black, center of bottom edge	not analyzed	retail trade paint: oil-modified alkyd[+]
Untitled, 1949	black, from reverse	sulfur	protein-based ink
Woman, 1953	orange, bottom left edge	not analyzed	drying oil + castor oil (P/S 1.2)[#]
	yellow, skirt, below proper right arm	not analyzed	drying oil + castor oil (P/S 1.0)[#]
	light green, near top right corner	not analyzed	drying oil + castor oil (P/S 0.7)[#]
	dark green, bottom left edge	not analyzed	drying oil + castor oil (P/S 0.7)[#]
Woman VI, 1953	black, near breasts	not analyzed	drying oil + castor oil (P/S 1.0)[0]
	pink, proper left arm	not analyzed	drying oil + castor oil (P/S 1.0)[0]
Two Women in the Country, 1954	black, right edge, center	not analyzed	retail trade paint: oil-modified alkyd[+]
	white, left edge, center	not analyzed	soybean or safflower oil (P/S 2.7)[#]
	orange, near right edge, center	not analyzed	linseed oil (P/S 1.7)[#]
Gotham News, 1955	black, near center	not analyzed	drying oil + castor oil (P/S 0.5)[0]
	white, upper left quadrant	not analyzed	drying oil + castor oil (P/S 0.5)[0]

P=palmitate; S=stearate

It is seldom possible to conclusively identify mixtures of different binding media. The media listed in these paintings reflect a cautious interpretation based on the peak area and molar ratios given and comparison of them with naturally aged paint standards, both neat and with mixed binding media.

[+] Results obtained from procedure described in Appendix 2B.

[0] Results obtained from procedure described in Appendix 2F.

[#] Results obtained from procedure described in Appendix 2D.

[*] Pyrolysis gas-chromatography. Sample reacted with trimetyl (trifluoro-m-toly)ammonium hydroxide to convert the pyrolyzates to methyl esters (Chris Maines).

[b] Minor amounts of pigment were noted.

TABLE 2 Analysis of de Kooning's Paints in Works from the 1960s

Painting	Sample Site	Pigments	Binding Medium
Door to the River, 1960	off-white, left of center	not analyzed	poppy oil (P/S 4.1)[o]
	pink, right of center	not analyzed	poppy oil + linseed oil (P/S 3.5)[o]
Spike's Folly II, 1960	blue, top center	not analyzed	poppy oil + castor oil (P/S 3.7)[o]
	yellow, near center of right edge	not analyzed	drying oil + castor oil (P/S 0.90)[o]
Untitled, May 1962	ground	titanium dioxide calcium sulfate clay	drying oil (P/S 2.1)[+]
	yellow, center of bottom edge	barium sulfate zinc oxide titanium dioxide cadmium yellow organic yellow[b]	poppy oil (P/S 4.4)[+]
	pink, left of center	barium sulfate titanium dioxide zinc oxide organic red	poppy oil (P/S 4.8)[+]
	gray-brown, center of left edge	cadmium orange cadmium yellow ultramarine blue (art.) organic red	retail trade paint: oil-modified alkyd[+]
Rosy-Fingered Dawn at Louse Point, 1963	yellow	not analyzed	linseed oil + castor oil (P/S 1.9)[o]
Pastorale, 1963	light blue, upper right quadrant	not analyzed	poppy oil (P/S 4.8)[o]
	gray-brown, near lower left corner	not analyzed	retail trade paint: oil-modified alkyd[o]
Two Standing Women, 1963–64	coated covering paper support	unpigmented	acrylic resin
	yellow, left edge	cadmium yellow	poppy oil (P/S 4.3)
	pink	not analyzed	poppy oil (P/S 4.1)
Sphinx, 1964	red, below signature	cadmium red [CdS(Se)]	safflower oil + linseed oil (P/S 2.3)[+]
	pink, center of bottom edge	barium sulfate cadmium orange zinc oxide titanium dioxide	safflower oil + linseed oil (P/S 2.9)[+]
Woman, 1964	coating covering paper support	unpigmented	synthetic resin (estimate)
	pink, lower right edge	titanium dioxide zinc oxide cadmium red barium sulfate	poppy oil (P/S 4.3)
	yellow, left edge	barium sulfate cadmium yellow titanium dioxide zinc oxide	poppy oil (P/S 4.5)
Woman, Sag Harbor, 1964	ground	zinc oxide titanium dioxide	drying oil (P/S 2.4)[+]
	white, lower left corner	barium sulfate titanium dioxide zinc oxide	safflower oil + linseed oil (P/S 2.5)[+]
	pink, woman's upper torso	titanium dioxide barium sulfate zinc oxide cadmium red organic red[b]	safflower oil + linseed oil (P/S 2.5)[+]

continued

TABLE 2 (CONTINUED)

Painting	Sample Site	Pigments	Binding Medium
	blue-green, right of center	titanium dioxide zinc oxide barium sulfate cadmium yellow organic green organic blue	safflower oil + linseed oil (P/S 2.8)[+]
	maroon from genitals	cadmium red [CdS(Se)]	safflower oil + linseed oil (P/S 2.5)[+]
	red, woman's proper right shoulder	cadmium red [CdS(Se)]	safflower oil + linseed oil (P/S 2.3)[+]
	yellow, left edge	zinc oxide cadmium yellow titanium dioxide[b]	safflower oil + linseed oil (P/S 3.3)[+]
	orange, from woman's hair	cadmium orange [CdS(Se)]	linseed oil (P/S 1.7)[+]
	light pink, between woman's shoes	not analyzed	safflower oil or poppy oil + linseed oil (P/S 3.4)[+]
Woman, 1965	orange, left of woman's torso	barium sulfate zinc oxide cadmium orange titanium dioxide organic red	safflower oil + linseed oil (P/S 2.9)[+]
	red, near woman's proper left leg	cadmium red [CdS(Se)] organic red	safflower oil + linseed oil (P/S 2.8)[+]
	dark green, left edge	barium sulfate cadmium yellow cadmium orange molybdate zinc oxide organic green ultramarine blue (art.) carbon black[b]	safflower oil + linseed oil (P/S 2.8)[+]
	light blue, left edge	barium sulfate zinc oxide titanium dioxide ultramarine blue (art.)	poppy oil + linseed oil (P/S 3.8)[+]
	pink, woman's lower torso	barium sulfate cadmium red cadmium orange molybdate zinc oxide	safflower oil + linseed oil (P/S 2.8)[+]
	pink, woman's upper torso	not analyzed	safflower oil + linseed oil (P/S 2.6)[+]
The Visit, 1966–67	orange, near center of left edge	not analyzed	safflower oil + linseed oil (P/S 2.6)[0]
	blue-gray, near bottom edge	not analyzed	safflower oil + linseed oil (P/S 2.6)[0]
Two Figures in the Landscape, 1967	pink, right of center	not analyzed	safflower oil + linseed oil (P/S 2.7)[0]
	red-orange, near center	not analyzed	safflower oil + linseed oil (P/S 2.1)[0]
	medium-rich looking pink, near center	not analyzed	safflower oil + linseed oil (P/S 2.8)[0]

P=palmitate; S=stearate

It is seldom possible to conclusively identify mixtures of different binding media. The media listed in these paintings reflect a cautious interpretation based on the peak area and molar ratios given and comparison of them with naturally aged paint standards, both neat and with mixed binding media.

[+] Results obtained from procedure described in Appendix 2B.

[0] Results obtained from procedure described in Appendix 2F.

[b] Minor amounts of pigment were noted.

TABLE 3 Analysis of de Kooning's Binding Media in Works from the 1970s

Painting	Sample Site	Binding Medium
Amityville, 1971	yellow, left edge	poppy oil + linseed oil (P/S 3.9)°
	white, left quadrant	poppy oil + linseed oil (P/S 3.7)°
. . . Whose Name Was Writ in Water, 1975	amber-colored drip, lower left quadrant	linseed oil + safflower oil (P/S 2.6)°
	red, lower left quadrant	safflower oil (P/S 2.7)°
	maroon, near center	poppy oil + castor oil (P/S 3.8)°
Untitled I, 1977	white, lower right quadrant	linseed oil (P/S 1.3)°
	pink, upper left quadrant	linseed oil (P/S 1.4)°
Untitled V, 1977	gray, upper left quadrant	linseed oil (P/S 1.8)°
	black, left edge, center	castor oil (P/S 0.91)°
	dark blue, lower right quadrant	linseed oil (P/S 2.0)°

P=palmitate; S=stearate

It is seldom possible to conclusively identify mixtures of different binding media. The media listed in these paintings reflect a cautious interpretation based on the peak area and molar ratios given and comparison of them with naturally aged paint standards, both neat and with mixed binding media.

° Results obtained from procedure described in Appendix 2F.

TABLE 4 Weight Percentages and Molar Ratios for Artists' Liquid Oils

	Weight Percentages								Molar Ratios	
Product	Glycerol	Palmitic acid C16 saturated	Stearic acid C18 saturated	Oleic acid C18:1 unsaturated	Linoleic acid C18:2 unsaturated	Linolenic acid C18:3 unsaturated	Ricinoleic acid 12-hydroxy-9-*cis*-C18:1	Eleostearic acid 9,11,13-C18:3	P/S	P/G
Linseed oil	9.7	4.8	3.2	17	10	55	0	0	1.7	0.18
Poppy oil	9.8	8.9	2.1	15	49	15	0	0	4.7	0.33
Safflower oil	9	5.2	2	14	55	15	0	0	2.9	0.21
Sunflower oil	9.7	5.7	4.4	20	56	4	0	0	1.4	0.21
Tung oil	10	2.2	2.1	10	14	0	0	62	1.1	0.08
Walnut oil	8.4	6.7	2.1	14	44	25	0	0	3.5	0.28
Castor oil	9.1	0.86	0.87	4	32	0	53	0	1.1	0.03

TABLE 5 Peak Area Ratios of Paint Samples (Relative to C_{17} Internal Standard)*

	Solvent Extract		Residue in Paint	
Paint Sample	FFA	DA	FFA	DA
MU1 – Cd orange tube color	0.05	0.007	0.45	0.17
MU8 – 1:1 mixture of MU1 and safflower oil	0.29	0.02	1.78	1.36
MU22 – 3:1 mixture of MU8 and water	0.91	0.04	1.89	1.36
Woman, Sag Harbor – hard white	0.31	0.02	2.13	1.10
Woman, Sag Harbor – gooey maroon	1.73	0.08	0.35	0.25
Woman (1965) – hard pink	0.58	0.01	2.10	1.21
Woman (1965) – soft orange	0.52	0.01	1.13	0.93

MU: Mock-up

FFA: total for all fatty acids

DA: total of C_8, C_9 and C_{10} dicarboxylic fatty acids

* The peak area ratios are the average of triplicate runs.

TABLE 6 Fatty Acid and Glycerol Results for Paints with Low, Mid-, and High Palmitate to Stearate (P/S) Peak Area Ratios

Fatty Acid and Glycerol Results for Paints with Low P/S Ratios

Painting	Date	Sample Description	% Oil	% FFA	P/S	P/G
Spike's Folly II	1960	Yellow, fairly soft, very matte	52	40	0.90	0.04
Untitled V	1977	Black, wrinkled	48	66	0.91	0.03
Untitled I	1977	White, wrinkled	59	23	1.3	0.05
Untitled I	1977	Pink-white, wrinkled, bumpy texture	67	37	1.4	0.04
Untitled V	1977	Gray, wrinkled, voids	40	17	1.8	0.07
Rosy-Fingered Dawn at Louse Point	1963	Bright yellow, gooey	50	48	1.9	0.07
Untitled V	1977	Dark blue, heavily wrinkled, very soft	59	37	2.0	0.07
		Average Values	54	38	–	–

Fatty Acid and Glycerol Results for Paints with Mid P/S Ratios

Painting	Date	Sample Description	% Oil	% FFA	P/S	P/G
Two Figures in a Landscape	1967	Orange, very soft, rubbery/gooey	31	29	2.1	0.17
The Visit	1966	Orange, gooey	24	33	2.6	0.12
Woman	1965	Amber material from drip on surface	52	17	2.6	0.14
...Whose Name Was Writ in Water	1975	Amber drip, very sticky, from surface	29	25	2.6	0.09
The Visit	1966	Wrinkled gray-green, soft	31	34	2.6	0.15
Two Figures in a Landscape	1967	Pink, wrinkled, soft	33	32	2.7	0.16
...Whose Name Was Writ in Water	1975	Red, very soft, sticky	67	26	2.7	0.14
Two Figures in a Landscape	1967	Amber gel, recess of wrinkled flesh	59	22	2.8	0.09
		Average Values	41	27	–	–

Fatty Acid and Glycerol Results for Paints with High P/S Ratios

Painting	Date	Sample Description	% Oil	% FFA	P/S	P/G
Woman, Sag Harbor	1964	Pink, soft	18	29	3.4	0.10
Door to the River	1960	Pink, voids, matte	25	32	3.5	0.26
Spike's Folly II	1960	Blue, very brittle, matte	23	12	3.7	0.24
Amityville	1971	White, soft	38	15	3.7	0.24
...Whose Name Was Writ in Water	1975	Maroon, very soft, stringy	24	26	3.8	0.13
Amityville	1971	Yellow, wrinkled, soft	23	28	3.9	0.21
Door to the River	1960	White, voids, matte, fairly brittle	29	19	4.1	0.26
Pastorale	1963	Light blue, voids, very brittle	48	32	4.8	0.20
		Average Values	29	24	–	–

Appendix 2

Methods of Sample Preparation

Susan F. Lake and Michael Schilling

A: Optical Microscopy (PLM)

All of the paintings included in this study were examined and photographed in conservation studios using a stereomicroscope for magnified study of the surface. Microscopic paint samples were taken from all of the paintings in the study. Both paint cross-sections, which include the layers of a painting, and dispersed pigment samples, which isolate a specific paint layer, were analyzed. Dispersed paint samples, consisting of a few particles of pigment from the paint surface, were mounted in resin for examination by polarized microscopy. This technique permitted the identification of specific pigments on the basis of optical properties. Microscopic paint cross-sections were taken from existing areas of paint loss or from the edges of the painting. The cross-sections were mounted in polyester resin and ground and polished for examination by reflected light microscopy. This technique permits the examination of the pigments in the context of the paint layer structure.

B: Oil Analysis Using Gas Chromatography/ Mass Spectrometry (GC/MS) with Meth-Prep II Procedure

The sample was placed in a 1-mL conical-bottomed reaction vial, to which 20 μL of a 5% solution of *m*-(trifluoromethyl)phenyl-trimethylammonium hydroxide in methanol (Meth Prep II reagent, from Grace Davison Discovery Sciences) was added. The vial was heated to 60°C for 4 hours, kept at room temperature overnight, then centrifuged prior to GC/MS analysis. A ZAB 2SE mass-spectrometer (Micromass, Manchester, U.K.) was used with an HP-5 column (30 M × 0.32 mm × 0.32 μM ID), splitless injection at 250°C, and helium carrier gas at 1 mL/minute. The oven temperature program was: 3 minutes at 100°C, 15°C/minute to 200°C, 10°C/minute to 300°C, and 5 minutes isothermal.

C: Oil Analysis Using GC with Saponification and Diazomethane Procedure

The sample was placed in a 1-mL conical vial, to which 0.2 mL of KOH solution (10% in methanol) was added, and the vial was heated to 90°C for 2 hours. After cooling to room temperature, 0.2 mL of water was added, and the solution was made acidic by dropwise addition of concentrated HCl (the acidity was determined with pH paper). The organic materials were removed by extracting twice with diethyl ether, the ether layers combined into a clean conical vial, and the ether evaporated. The sample was then dissolved in 0.5 mL of 10% methanol in diethyl ether. The sample was reacted with diazomethane (Schlenk and Gellerman 1960), and the contents of the vial were then concentrated down for analysis of the fatty acid methyl esters by gas chromatography (GC).

The analyses were performed on a Perkin Elmer Autosystem GC equipped with a flame ionization detector, a Quadrex Al-coated methyl silicone capillary column (25 M × 0.25 mm × 0.25 μM), splitless inlet at 300°C with a 30-second purge delay, helium carrier at 9.0 psig, and the detector at 325°C. The oven temperature program was: 5 minutes at 130°C, 4°C/minute to 300°C, and 20 minutes isothermal. Standards of dimethyl azelate, methyl palmitate, and methyl stearate were injected for comparison with the unknown materials.

D: Pyrolysis-Gas Chromatography/ Mass Spectrometry (Pyrolysis-GC/MS)

A Frontier Lab PY-2020D double-shot pyrolyzer system equipped with a CGS-1050E carrier gas selector, SS-1010E selective sampler, and MJT-1030E microjet cryo-trap was used for pyrolysis. The pyrolysis interface was maintained at 320°C. The pyrolyzer was interfaced to an Agilent Technologies 5975C inert MSD/7890A GC/MS. A Frontier Ultra ALLOY-5 capillary column was used for the separation (30 M × 0.25 mm × 0.25 μm), with helium carrier gas set to 1 mL/minute. The split injector was set to 320°C with a split ratio of 50:1 and no solvent delay. The GC oven temperature program was: 2 minutes at 40°C, 20°C/minute to 320°C, and 9 minutes isothermal. The MS transfer line was at 320°C, the source at 230°C, and the MS quad at 150°C. The mass spectrometer was scanned from 33–600 amu at a rate of 2.59 scans per second. The electron multiplier was set to the autotune value.

Samples were placed into a 50-μL stainless steel Eco-cup fitted with an Eco-stick, and 3 μL of 25% tetramethylammonium hydroxide (TMAH) in methanol was introduced for derivatization. After 3 minutes, the cup was placed into the pyrolysis interface where it was purged with helium for 3 minutes. Samples were pyrolyzed using a single-shot method at 550°C for 6 seconds.

E: Solvent Extraction Procedure

The sample was weighed (roughly 300 μg), transferred to a 1-mL conical vial, and a 1:1 mixture of benzene and hexane was added to extract the components not bound in the cross-linked network of the paint film. After standing overnight at room temperature, the supernatant was transferred to a second vial and evaporated to dryness using a nitrogen stream. Trimethylsilyl esters of the free fatty acids were formed by the addition of Tri-Sil reagent (Thermo Fisher Scientific). The residual paint sample was subsequently derivatized and methylated for GC/MS analysis using a Meth-Prep II reagent. A 100-ppm internal standard of heptadecanoic acid was added to each of the samples and the GC/MS procedure in Appendix 2B was used. The quantitative results were expressed as a ratio of the fatty acids to the internal standard.

F: Total Fatty Acid and Glycerol Analysis

Methanolic base (approximately 0.5M sodium methoxide in methanol) and methanolic HCl (approximately 1.5M HCl in methanol) were obtained from Sigma-Aldrich. Prior to use, the methanolic base strength was checked by titration against potassium hydrogen phthalate (Sigma-Aldrich) to a red phenolphthalein endpoint. Methanolic acid strength was checked by back-titrating a known amount of excess base with acid to a colorless endpoint. Dilutions of acid and base for use in titrations were made with HPLC-grade water (Honeywell Burdick and Jackson).

A derivatizing solution consisting of a 5:2:3 ratio by volume of benzene (residue-grade, Mallinckrodt Baker), 2,2-dimethoxypropane (Sigma-Aldrich), and methanolic base was prepared in a 100-mL volumetric flask. Hexachlorobenzene (Sigma-Aldrich) was added to give a final concentration of approximately 100 ppm. A solid neutralizing mixture was prepared from a 2:1:2 ratio by weight of sodium bicarbonate, sodium carbonate, and anhydrous sodium sulfate (all from Merck).

Paint samples were weighed onto aluminum pans to the nearest 0.1 µg, transferred to 1-mL conical vials, and approximately 1 µL of reagents was added per µg of paint. The vials were shaken to promote extraction of the lipid materials, and afterwards were allowed to stand overnight. The following morning sufficient methanolic acid was added to the sample vials to neutralize the base and provide exactly 150.0 µmoles of excess acid per mL of reagents. The vials were shaken thoroughly, and the contents were allowed to react for exactly 60 minutes. Afterward, approximately 75 mg of solid neutralizer per mL of reagents was added to the vials, and after 30 minutes the solutions were ready for gas-chromatography–flame ionization detection (GC/FID) analysis.

The analyses were performed on a Hewlett Packard 6890 GC/FID using an HP-5MS column (30 M x 0.25 mm x 0.25 µm), helium carrier gas at 39 cm/second, 300°C splitless inlet (50 mL/min vent flow, 1 mL/min septum purge, 60 second purge delay), FID at 300°C, 20 Hz data collection rate. The oven temperature program was: 2 minutes at 50°C, then 20°C/minute to 300°C.

Standards of fatty acid methyl esters (FAMEs) and isopropylidene glycerol (IPG or Solketal) were used without further purification (all from Sigma-Aldrich). Calibration standards were prepared by serial dilution of a FAME and IPG stock solution; nominal concentrations ranged from 2 ppm to 100 ppm. Hexachlorobenzene was added as an internal standard at a concentration of 100 ppm.

G: Analysis of Free Fatty Acids as Trimethylsilyl Esters

Paint samples weighed onto aluminum pans to the nearest 0.1 µg were transferred to 1-mL conical vials, and Tri-Sil reagent was added (approximately 1 µL of reagent per µg of paint). GC/MS analyses were performed on a Hewlett Packard 6890/5972, using an HP-5MS column (30 M × 0.25 mm × 0.25 µm), helium carrier gas at 46 cm/second, splitless injection at 320°C (50 mL/minute vent flow, 1 mL/minute septum purge, 60 second purge delay), and the transfer line was set to 300°C. The oven temperature program was: 1 minute at 80°C, 20°C/minute to 320°C, and 15 minutes isothermal. The ions selected for quantitation were the molecular peaks of the trimethylsilyl esters.

Standards of saturated free fatty acids and free dicarboxylic fatty acids (Sigma-Aldrich) were tested to permit quantitative analysis. Nominal concentrations ranged from 2 ppm to 100 ppm. Hexachlorobenzene was added as an internal standard at a concentration of 100 ppm.

H: Scanning Electron Microscopy (SEM) with Energy Dispersive Spectroscopy (EDS)

Qualitative EDS was done on paint samples, some mounted in a polyester resin block as cross-sections and some unmounted, to determine the major and minor elements present. The samples were placed onto a carbon SEM stub with conductive graphite adhesive. The sample stub was then coated with a thin layer of evaporated graphite to increase conductivity under the electron beam. The instrument used for analysis of the paint samples in cross-section was a JEOL JXA-840A scanning electron microscope with a Tracor Northern TN-5502 energy dispersive x-ray spectroscopy system with a low atomic number detector (Pioneer Premium detector with a Norvar window). These spectra were obtained at 20 kV for a 60-second live-time acquisition. The unmounted paint samples were examined with a Leo S91B scanning electron microscope with a PGT (Princeton Gamma Technology) energy dispersive x-ray spectroscopy system. Analysis was done at 15 kV. In both forms of analysis, elemental spectra were acquired for at least three general areas of each paint color.

I: Fourier Transform Infrared Spectroscopy

Microsamples of paint or resin were placed neat into a diamond anvil cell and compressed to form a thin film. The instrument used was a Mattson 4326 Upgrade Fourier-transform infrared spectrophotometer equipped with a Spectra Tech IR Plan microscope. The infrared spectra were collected between 4000 and 600 cm^{-1} at 8 cm^{-1} resolution in 3 minutes scanning time (215 scans).

J: X-ray Diffraction (XRD)

For some samples, XRD provided structural confirmation of the pigments' identities; this method did not work for barium pigments, due to interferences with intense diffraction lines of these pigments. A Philips XRG 3100 x-ray generator was used, operated at 45 kV and 25 mA. The samples were adhered onto glass fibers with silicone grease and placed in Gandolfi cameras along with strips of Kodak DEF 392 Direct Exposure film. The samples were exposed for 6 hours in air. The film-strips were developed and fixed using Kodak GBX developer and fixer. Lines were read and intensities estimated by hand. Data were compared to standards run in the laboratory or to the Powder Diffraction File (1997; www.icdd.com).

Appendix 3

Works of Art Examined

Seated Woman, 1943

Graphite on wove paper
33.7 × 26 cm (13¼ × 10¼ in.)
Philadelphia Museum of Art
Bequest of Elizabeth M. Petrie, 2003

Head of a Woman (Study for Queen of Hearts), ca. 1943–45

Pen and wash on brown wove paper (paper towel)
28 × 20 cm (11 × 7⅞ in.)
Iris & B. Gerald Cantor Center for Visual Arts at Stanford University
Gift of Victor and Paula Zurcher

Queen of Hearts, 1943–46

Oil and charcoal on fiberboard
117 × 70 cm (46⅛ × 27⅝ in.)
Hirshhorn Museum and Sculpture Garden, Smithsonian Institution
Gift of the Joseph H. Hirshhorn Foundation, 1966

Special Delivery, 1946

Oil and charcoal on paper mounted on fiberboard
59.3 × 76.1 cm (23⅛ × 30 in.)
Hirshhorn Museum and Sculpture Garden, Smithsonian Institution
Gift of the Joseph H. Hirshhorn Foundation, 1966

Valentine, 1947

Oil on paper on board
92.2 × 61.5 cm (36⅜ × 24¼ in.)
Museum of Modern Art, New York
Gift of Mr. and Mrs. Gifford Phillips

Zurich, 1947

Oil on paper, mounted on fiberboard
91.4 × 61.3 cm (36 × 24⅛ in.)
Hirshhorn Museum and Sculpture Garden, Smithsonian Institution
Joseph H. Hirshhorn Bequest, 1981

Untitled, 1947–48

Oil on paper
50.8 × 40.6 cm (20 × 16 in.)
Allan Stone Gallery, New York

Black Friday, 1948

Oil on fiberboard
125.1 × 99.1 cm (49½ × 39 in.)
The Art Museum, Princeton University
Gift of H. Gates Lloyd, Class of 1923, and Mrs. Lloyd, in honor of the Class of 1923

Secretary, 1948

Oil and charcoal on paper mounted on fiberboard
62 × 92 cm (24⅜ × 36 3⁄16 in.)
Hirshhorn Museum and Sculpture Garden, Smithsonian Institution
Gift of Joseph H. Hirshhorn, 1966

Woman, 1948

Oil on fiberboard
136.1 × 113.2 cm (53⅝ × 44⅝ in.)
Hirshhorn Museum and Sculpture Garden, Smithsonian Institution
Gift of the Joseph H. Hirshhorn Foundation, 1966

Untitled, 1949

Ink on paper
55.7 × 76.2 cm (21 11⁄16 × 30 in.)
Hirshhorn Museum and Sculpture Garden, Smithsonian Institution
Gift of Joseph H. Hirshhorn, 1966

Woman, 1953

Oil and charcoal on paper, mounted on canvas
64.9 × 49.8 cm (25⅝ × 19⅝ in.)
Hirshhorn Museum and Sculpture Garden, Smithsonian Institution
Gift of Joseph H. Hirshhorn, 1966

Woman VI, 1953

Oil on canvas
174 × 148.6 cm (68½ × 58½ in.)
Carnegie Museum of Art, Pittsburgh
Gift of G. David Thompson, 1955

Two Women in the Country, 1954

Oil and charcoal on canvas
117.1 × 103.5 cm (46⅛ × 40¾ in.)
Hirshhorn Museum and Sculpture Garden, Smithsonian Institution
Gift of Joseph H. Hirshhorn, 1966

Gotham News, 1955

Oil, charcoal, and newspaper ink transfer on canvas
175.3 × 200.7 cm (69 × 79 in.)
Albright-Knox Art Gallery, Buffalo, New York
Gift of Seymour H. Knox, 1955

Door to the River, 1960

Oil on canvas
203.2 × 177.8 cm (80 × 70 in.)
Whitney Museum of American Art, New York
Purchased with funds from the Friends of the Whitney Museum of American Art

Spike's Folly II, 1960

Oil on canvas
203.2 × 177.8 cm (80 × 70 in.)
Robert and Jan Meyerhoff, Phoenix, Md.

Untitled, May 1962

Oil on canvas
202.9 × 179.5 cm (79⅞ × 70⅝ in.)
Hirshhorn Museum and Sculpture Garden, Smithsonian Institution
Gift of the Joseph H. Hirshhorn Foundation, 1966

Pastorale, 1963

Oil on canvas
177.8 × 203.2 cm (70 × 80 in.)
Private Collection, New Orleans

Rosy-Fingered Dawn at Louse Point, 1963

Oil on canvas
203.2 × 177.8 cm (80 × 70 in.)
Stedelijk Museum, Amsterdam

Two Standing Women, 1963–64

Oil and charcoal on paper, mounted on canvas
73.7 × 58.8 cm (29 1/8 × 23 1/4 in.)
Hirshhorn Museum and Sculpture Garden, Smithsonian Institution
Gift of Joseph H. Hirshhorn, 1966

Sphinx, 1964

Oil and charcoal on paper
60.5 × 47.3 cm (23 13/16 × 18 5/8 in.)
Hirshhorn Museum and Sculpture Garden, Smithsonian Institution
Gift of Joseph H. Hirshhorn, 1966

Woman, 1964

Oil and charcoal, mounted on fiberboard
73.7 × 58.4 cm (29 × 23 in.)
Hirshhorn Museum and Sculpture Garden, Smithsonian Institution
Gift of Joseph H. Hirshhorn, 1966

Woman, 1964

Oil on vellum paper mounted on paperboard
193.8 × 96.4 cm (76 3/8 × 36 in.)
Hirshhorn Museum and Sculpture Garden, Smithsonian Institution
Gift of Joseph H. Hirshhorn, 1966

Woman, Sag Harbor, 1964

Oil and charcoal on wood
203.1 × 91.2 cm (80 × 36 in.)
Hirshhorn Museum and Sculpture Garden, Smithsonian Institution
Gift of Joseph H. Hirshhorn, 1966

Woman, 1965

Oil on wood
203.2 × 91.4 cm (80 × 36 in.)
Hirshhorn Museum and Sculpture Garden, Smithsonian Institution
Gift of Joseph H. Hirshhorn, 1966

Untitled, 1966

Charcoal on paper
43.1 × 35.6 cm (17 × 14 1/16 in.)
Hirshhorn Museum and Sculpture Garden, Smithsonian Institution
Joseph H. Hirshhorn Bequest, 1981

The Visit, 1966–67

Oil on canvas
152.4 × 121.9 cm (60 × 48 in.)
Tate Gallery, London
Purchased 1969

Two Figures in a Landscape, 1967

Oil on canvas
177.8 × 203.2 cm (70 × 80 in.)
Stedelijk Museum, Amsterdam

Amityville, 1971

Oil on canvas
203.2 × 177.8 cm (80 × 70 in.)
Private collection

Landscape at Stanton Street, 1971

Lithograph drawn on stone.
Black ink on Dutch etching paper
76.2 × 55.9 cm (30 × 22 in.)
Published by Hollanders Workshop and Fourcade, New York, 1971

...Whose Name Was Writ in Water, 1975

Oil on canvas
194.9 × 223 cm (76 3/4 × 87 3/4 in.)
Solomon R. Guggenheim Museum, New York
By exchange, 1980

Untitled I, 1977

Oil on canvas
195.6 × 223.5 cm (77 × 88 in.)
Adriana and Robert Mnuchin

Untitled V, 1977

Oil on canvas
201.9 × 175.9 cm (79 1/2 × 69 1/4 in.)
Albright-Knox Art Gallery, Buffalo, New York
Gift of Seymour H. Knox, 1977

Notes

Introduction

1 As the late 1940s, 1950s, and 1960s were signature periods in de Kooning's career, these years are particularly appropriate for this study. It must be remembered, however, that these seminal decades lie within a six-decade-long artistic evolution marked by repeated transitions from figuration to abstract modes.

Chapter 1

2 Also see Lerner (1975). In 1983, de Kooning told an interviewer that he had not originally intended to become an artist but had come to the United States because American illustrators were so good at commercial layout. He also said that he thought the girls looked fantastic. He liked western movies, jazz, and Louis Armstrong; see Pepper (1983, 86).

3 Dore Ashton (1969, 244) observed that the black-and-white paintings, executed from 1947 through 1950, were the last works by de Kooning to be universally admired.

4 De Kooning reportedly noted that it was Pollock who "broke the ice" (Sandler 1978). In a 1967 interview, de Kooning said that "[Pollock] told me 'You know more, but I feel more.' I was jealous of him—of his talent. But he was a remarkable person. He'd do things that were so terrific" (Vallière 1967, 603–5). It should be noted that at the same time, Hans Hofmann was also expanding his use of "accidental" painterly effects, Franz Kline was applying his paints with greater speed, and Clyfford Still was creating his increasingly abstract compositions with a palette knife.

5 In addition to de Kooning, the photograph included artists Theodoros Stamos, Jimmy Ernst, Barnett Newman, James Brooks, Mark Rothko, Richard Pousette-Dart, William Baziotes, Jackson Pollock, Clyfford Still, Robert Motherwell, Bradley Walker Tomlin, Adolph Gottlieb, Ad Reinhardt, and Hedda Sterne.

6 More than ten years later de Kooning also noted, "I was influenced by lots of artists. If you paint your whole life … after a while all kinds of painting become just painting for you—abstract or otherwise" (*Time* 1951, 63).

7 For a thorough examination of the critical responses to de Kooning's women paintings, see Cateforis (1992).

8 From the early 1950s onward, de Kooning and Elaine frequently spent portions of the summer with fellow artists and dealers in East Hampton. In the summer of 1961, the artist bought a small house in the Springs area of East Hampton.

9 Also reprinted in T. Hess (1968, 141–43).

10 De Kooning said of Still and Pollock: "They stand all alone in the wilderness—breast bared. This is an American idea.… I am a foreigner after all. I am different from them because I am interested in all of art," quoted in Sandler (1989, 216).

11 Hess's 1953 article was one of a regular series that *ARTnews* began in 1949 on a wide range of contemporary artists.

12 In its 1959 Summer and September issues, *ARTnews* published the results of a two-part symposium, "Is There a New Academy?" In 1967, Hilton Kramer noted that the second generation's fabrications after de Kooning had become a "Tenth Street Industry"; see Kramer 1973. In his review of de Kooning's retrospective at the Tate Gallery, London, critic David Thompson (1969) described the artist as part of a now-irrelevant historical style. Robert Goldwater, in "Reflections on the New York School" (1960), examined the revisionist mood at this time. For an overview of the changing critical attitudes in the 1950s and 1960s, see Rosenzweig (1980) and Kleeblatt (2008).

13 See, for example, *De Kooning: Paintings, 1960–1980* (Basel: Kunstmuseum, 2005); *Willem de Kooning: Tracing the Figure* (Los Angeles Museum of Contemporary Art, 2002); *Willem de Kooning: In Process* (Fort Lauderdale: Museum of Art, 2000); *Willem de Kooning's Door Cycle* (New York: Whitney Museum of American Art, 1996); *Willem de Kooning: The Late Paintings, the 1980s* (San Francisco and Minneapolis: Museum of Modern Art and Walker Art Center, 1995); *Willem de Kooning: Paintings* (Washington, D.C.: National Gallery of Art, 1994).

14 T. Hess, *Willem de Kooning: Drawings* (1972c, 13–17), contains a concise description of de Kooning's working methods.

15 John Brealey, chairman of the Department of Paintings Conservation at the Metropolitan Museum of Art, New York, claimed that the color values in *Easter Monday*, 1955–56, have altered to an "ugly rust" due to the discoloration of the whites, the result of the artist's addition of salad oil to his paints.

16 Interdepartmental Memorandum, from John Brealey to Thomas Hess, June 28, 1978, Curatorial Records, Department of Twentieth-Century of Art, Metropolitan Museum of Art.

Chapter 2

17 Rather than using the generic term "commercial" paint, which often is used to denote non-art paints, this study uses instead the terminology adopted by Marion Barclay in her essay, "Materials Used in Certain Canadian Abstract Paintings of the 1950s" (1992). "Retail trade paints" refer to products that are designed for interior

and exterior use to coat architectural structures, signs, and furniture. They can be readily purchased by the general public. Industrial paints are products used in the manufacturing process, such as automotive paints or stoving enamels. These paints can only be bought with some difficulty by the general public.

18 Masonite is a fiber or particle board first produced in 1924 by the Masonite Corporation. Although fiberboard was produced by many companies, the term "Masonite" was applied generically to many types and kinds of fiberboard. The ground layer was identified by microchemical analysis and inferences from SEM-EDS. A small amount of titanium dioxide is present in the gesso. The protein-based glue was identified with FTIR spectroscopy by comparing a water extract of the ground binder with that of hide glue.

19 Shellac was identified based on the presence of a faint orange fluorescent layer visible in ultraviolet light, located immediately above the ground in several of the paint cross-sections.

20 Two cross-sections taken from the upper right and lower right corners show residues of blue paint below the ground, indicating that the artist scraped down an earlier composition prior to applying the current priming layer.

21 Twenty samples of pigmented and unpigmented linseed oil run by the author gave an average palmitate-to-stearate (P/S) peak area ratio of 1.7 with a range from 1.4 to 1.8.

22 Extender and filler pigments are those inert white pigments that do not function primarily as coloring agents due to refractive indices below 1.7.

23 De Kooning often expressed his admiration for sign painters, who used a repertoire of trade-secret techniques and quick precision to create bold, direct visual messages.

24 Natural resin was identified by FTIR spectroscopy.

25 Both examination methods indicate that the artist first painted the background areas around an empty space that he reserved for the figure. Nonetheless, given de Kooning's lifelong practice of reworking his images, it seems likely that he painted the current figure of the female after he removed a previous image, or images, by scraping the central area of the support down to the ground and starting anew. Such revisions to the figure, in addition to the slow build-up of paint layers in the background, would account for the three years it took to complete this painting.

26 According to Thomas Hess, de Kooning, like many of the Abstract Expressionist painters, was reacting against the then-dominant aesthetic of the School of Paris, with its emphasis on balance and refinement. Painters like de Kooning and Pollock, Hess said, "omitted the last stages of 'finish.' The picture exposed the obstacles and how the artist had tried to overcome them" (T. Hess 1968, 23). For more on the twentieth-century move from mimetic rendition of symbols to indexial registering of traces, see Shiff (1987).

27 Thomas Hess reinforced this interpretation of the artist's working method in his essay, "De Kooning Paints a Picture" (T. Hess 1953).

28 In his classes, artist Hans Hofmann encouraged similar exercises of cutting and reassembling drawings as a means of freeing the students' imaginations; see Rose (1978, 114, n. 9).

29 The liner's brush, traditionally used by sign and theater scene painters, was one of de Kooning's favorite brushes. He was introduced to it as a student and continued to use it as a sign painter in the 1930s; see T. Hess (1959b, 23, and 1972c, 17). De Kooning introduced Gorky to this brush and it proved to be a tool essential to his painting; see Seitz (1962, 34).

30 Beginning in the 1930s, de Kooning studied his paintings and drawings from different angles and by the mid-1940s he had begun to paint from all sides, turning the support as he worked; see Yard (1986, 149). De Kooning seldom signed his paintings until just before they left his studio, and a completed painting was often signed only after he studied it from different orientations; see T. Hess (1959b, 14). The artist continued this practice well into the 1980s; see Garrels (1995, 22).

31 The identification of calcium sulfate was based on microchemical tests, polarized-light microscopy of dispersed samples, and inferences from SEM-EDS analysis. An FTIR spectrum indicates that the calcium sulfate is the hemi-hydrate form, or plaster of Paris ($CaSO_4 \cdot 1/2\ H_2O$).

32 The identity of the pigments in the yellow retail trade paint—cadmium yellow, titanium white, and barium sulfate—is based on inferences from SEM-EDS analysis.

33 Although de Kooning may have added the colophony to his artist's paints, this would have required heating the resin until it became liquid, an arduous procedure. It is more likely that he bought this paint as a prepared house or sign paint. Although it was not recommended, colophony frequently was an adulterant used in the production of cheaper grades of oil paints.

34 According to Thomas Hess, de Kooning was concerned that the colored pigments in house paints would fade, so when he used colored paints, even in the black-and-white abstractions, they were artists' colors; see T. Hess (1959b, 116).

35 Due to its low refractive index, calcium sulfate is not used neat in oils to make paints, although it sometimes is used as an additive or extender in oil paint systems; see Patton (1973, 290).

36 The varnish was prepared by Melvin Wachowiak, assistant director for conservation at the Smithsonian Museum Conservation Institute, following a nineteenth-century recipe for furniture varnish.

37 In 1946, Rainbow products marketed an extensive line of dry materials that included pumice, rottenstone, and dry pigments, in addition to plaster of Paris, according to a telephone conversation, September 10, 1998, with Jay Gornitzky, President, Empire Blended Products, Bayville, N.J.

38 The "greasy" appearance of the yellow paint varies greatly across the painting. The analyzed sample was taken from one of the smoother, more waxy-looking areas. The determination that this paint is an artist's paint is based on the identification of the binding medium as linseed oil with a P/S ratio of 1.9 and cadmium yellow as the sole pigment, with none of the additives and extenders usually associated with retail trade paints.

39 See Martens (1974, 546). Identification of the pigments and the clay was based on polarized light microscopy of dispersed samples and inferences from SEM-EDS analysis. Oil-modified alkyd house paint was first introduced in the 1930s but did not become readily available until the 1940s; see *Ullman's Encyclopedia of Industrial Chemistry* 1985, c.v. "alkyd resins." A long-oil alkyd resin suitable for mixing with artists' oil colors was first marketed, by Winsor and Newton, only in 1958. Artists' tube paints employing an oil-modified alkyd medium were first marketed in 1975; see Winsor and Newton Artists' Alkyd Colours, Leaflet no. AL1 (Zora Pinney Collection, The Artists' Materials Research and Study Center, National Gallery of Art, Washington, D.C.).

40 The quartz was differentiated from the glass based on its birefringence under crossed-polar illumination. De Kooning occasionally added pumice to his black paint to dull its gloss; see T. Hess (1972c, 49).

41 Analysis was done using a combination of polarized light microscopy and inferences from SEM-EDS analysis.

42 De Kooning began to include glossy black paints in his works around 1945, but he restricted their use to fluid drawing elements, often combined with charcoal, that supplemented his colored paints.

43 Reportedly, the color values of *Easter Monday* (1955–56) had "reversed themselves" due to the discoloration of the oil medium;

see Tomkins (1987, 66). The *Zurich* paint cross-section also shows an intense green layer above the lower layer of black, a reminder that many of the so-called black-and-white paintings contain patches of brilliant color.

44 Identification of the pigments was based on polarized light microscopy of dispersed samples and inferences from SEM-EDS analysis.

45 T. Hess (1968, 50) identified de Kooning's black enamel paint as Ripolin. In *Willem de Kooning: Drawings* (1972c, 13), he identified it as Sapolin. Janis and Blesh (1960, 26) identified the artist's black and white paints as Sapolin. Sapolin was a Manhattan company that manufactured oil paint, enamels, varnishes, and stains until it went out of business in the late 1970s; see *Thomas Register of American Manufacturers*, 1950–70.

46 The ratio of the sum of the areas of the abietic acids to that of methyl palmitate is about 1:1. The pigment identification was made from polarized light microscopy of dispersed samples and inferences from SEM-EDS analysis. The white pigment in *Valentine* and *Untitled* (1947–48) also is titanium dioxide. In retail trade paints, zinc oxide is often combined with titanium white, as the latter is a poor drier in oil. The addition of zinc oxide produces a harder film with more gloss; see Patton (1973, 42).

47 See Heaton (1947, 299, 350), and Brady and Clauser (1977, 662). Oil-ester gums were made with colophony heated to a molten state and converted to ester gum by the addition of 10–25% of its weight of glycerine. The heated mixture was then combined with oil, usually tung oil or dehydrated castor oil and driers. The relatively high amounts of the C_{14} methyl ester in the samples from both *Zurich* and *Black Friday* indicate that coconut oil may be present, in combination with another oil. The esterified gum paints were sometimes combined with cellulose nitrate resin. FTIR spectra of solvent extracts of samples from the black paints of both paintings indicate that cellulose nitrate is not present in either of these paints.

48 Quoted in "Inner Monologue" (1959, 11). De Kooning repeated this idea in 1964; see Dickerson (1964a, 70). In 1972, he again said that he had painted in black and white because he "needed a lot of paint and wanted to be free of the materials"; see Rosenberg (1972, 57).

49 Thomas Hess also frequently stressed that de Kooning's method was one that exploited chance and accident; see, for example, T. Hess (1972c, 43).

50 See Yard 1991 for an extended discussion of *Black Friday*. De Kooning often stated that he was never particularly concerned with the distinction between abstraction and figuration; see n. 6.

51 Six photographs by Rudolph Burckhardt of *Woman I* during different stages of its development were included in Thomas Hess's *ARTnews* article, "De Kooning Paints a Picture" (1953). Prather (1994, 128) also noted the connection between *Woman* (1948) and Burckhardt's stage-one photograph of *Woman I* in progress.

52 The intense pink paint of the woman's skirt is also an oleoresinous paint rather than an artist's oil, further confirmation that de Kooning's use of retail trade paints was not confined to white and black.

53 Plaster of Paris or calcium sulfate was identified based on microchemical tests, in addition to polarized light microscopy, inferences from SEM-EDS, and FTIR analysis.

54 Until around the 1920s, the choice of binders in retail trade paints was not dissimilar to those available to artists—drying oils mixed with a variety of natural resins—although they were admittedly of lesser quality. This formula had not altered significantly since the eighteenth century. By the 1940s, however, a whole range of synthetic and semi-synthetic paints were available, many of them by-products of the war industry, and these paints had begun to replace the traditional oils and varnishes in decorative paint formulations. Thus, for the first time in history, the paints available to industry and to the general public were significantly different from those sold in art stores.

55 *Ullman's Encyclopedia* (1985), c.v. "alkyd resins." Although the drying time of oil-modified alkyds varies with their oil length and the drying conditions, most house paints are designed to dry to the touch within four to six hours and they become insoluble—that is, sufficiently hardened to allow a further coat to be applied without disturbing the lower layer—within eighteen hours.

56 Through most of the 1940s, Sapolin enamel was a pigmented oleoresinous paint. By the late 1940s and early 1950s, however, Sapolin varnishes and enamels were quick drying, medium-length soya alkyd paints, according to Ted Friedman, a technical consultant for Insel-X Products, Yonkers, N.Y., which bought the Sapolin label after the company went out of business in the 1970s (Friedman 1992). The formulations for Ronan's drop black lettering enamel also changed in the late 1940s from oils combined with a variety of resins to an oil-modified alkyd, according to Morris Williams, technical director, T.J. Ronan Paint Corp., Bronx, N.Y. (Williams 1998).

57 Lacquer, like enamel, is a broadly used term applied to paints or varnishes that dry to a high gloss. De Kooning's use of the word reflects its modern commercial usage, where it has come to indicate a clear or pigmented coating material usually made from modified resins or derivatives of cellulose.

58 Tworkov (1950, 33) asserted in his review of the MoMA show that Soutine's work presaged certain attitudes toward paint on the part of the Abstract Expressionists with his "completely impulsive use of pigment as a material, generally thick, slow-flowing and viscous."

59 Pentaerythritol is one of the polyfunctional alcohols used in the manufacture of oil-modified alkyd resins.

60 McMahon became acquainted with de Kooning in the late 1950s and became his studio assistant from the middle of 1963 to 1973 or 1974. Although de Kooning was not officially represented by the Sidney Janis Gallery until 1953, the time of his first show there, they probably had arrived at an informal agreement as early as 1951. See also Stevens and Swan (2004, 319).

61 Two undated sets of Bocour's paint formulations are located in The Artists' Materials Research and Study Center, National Gallery of Art, Washington, D.C. Bocour brand Titanium White called for 1.5 parts castor oil and 1.0 parts linseed or safflower oil. While there was not sufficient sample to analyze further the paints from *Woman VI* and *Gotham News*, free fatty acid analysis of several samples from de Kooning's paintings of the 1960s, using a silylating reagent, confirmed the presence of ricinoleic acid, the principal acid in castor oil; see Chapter 4. The P/S peak area ratios of the binding media of the orange and white paints of *Two Women in the Country* are likely linseed oil and safflower or sunflower oil, respectively. By the 1950s, artists' paint manufacturers were more frequently grinding white pigments with safflower or sunflower oils because both oils tend to yellow less with time than linseed oil; see Gamblin (1998).

62 From Edwin Denby's poem "The Silence of Night," subtitled, "The designs on the sidewalk Bill pointed out."

Chapter 3

63 Several exhibitions have focused on de Kooning's work after his move from Manhattan: *Willem de Kooning in East Hampton* (New York: Solomon R. Guggenheim Museum, 1978); *De Kooning, 1969–1978* (Cedar Falls: University of Northern Iowa Gallery of Art, 1978); *Willem de Kooning: The North Atlantic Light* (Amsterdam:

Stedelijk Museum, 1983); *De Kooning: Recent Paintings* (New York: M. Knoedler and Company, 1967); *Willem de Kooning: Transcending Landscape: Paintings 1975–1979* (New York: C & M Arts, 1993); *Willem de Kooning: Paintings 1967–1987* (San Francisco: John Berggruen Gallery, 2002); and *De Kooning: Paintings 1960–1980* (Basel: Kunstmuseum, 2005). The two essays in *Willem de Kooning from the Hirshhorn Museum Collection* by Judith Zilczer and Lynne Cooke (1993) discuss de Kooning's career when he lived in New York and from 1963, when he moved to Long Island.

64 Thomas Hess was the first to characterize the paintings of the 1960s as pastoral, a mode he said still retained the perspective of the city; see T. Hess (1967, 9). More recently, Lynne Cooke examined de Kooning's East Hampton work in the historical context of the pastoral tradition and in the social context of the 1960s, in "De Kooning and the Pastoral: The Interrupted Idyll" (Cooke 1993).

65 See also Shiff (1994, 63 n. 5); Cowart and Shaman (1978, 15).

66 In another interview of the same year, de Kooning acknowledged that his work had changed in response to his new environment and added, "Yes, I have changed... I'm getting more interested in my own miracle in nature... and I don't want to grasp the abstract, not the reason, but the experience of having moments"; see "Inner Monologue" (1959, 8). In 1972, de Kooning told Harold Rosenberg that his move to the country had changed his work enormously (Rosenberg 1972, 57).

67 The size and ground layers were probably applied by the artist. Animal-glue size is used to seal the canvas fibers and prevent direct contact between the fabric and the oil layer that was to follow. The degradative effects of oils on fabric were well known, and a preliminary isolating size layer was a standard practice among artists. Anhydrous calcium sulfate precipitated with rutile titanium dioxide is a pigment mixture used in retail trade paints produced in this country from 1948 to 1972. Both calcium sulfate and kaolin were added to titanium dioxide to increase the opacity and leveling characteristics of the paint system; see Patton (1973, 15–16, 206).

68 De Kooning's emphatic, full-armed brushstrokes in these landscapes were often perceived, as were the earlier paintings of women, as arenas of violent activity. Sidney Janis reported that de Kooning's paintings of this period arrived at the gallery "punched through with ragged holes [where] the artist had rammed his brush through the canvas"; see Gaugh (1983, 59). The author has found no evidence of such structural damage in any of the paintings examined.

69 In a 1959 interview, de Kooning mentioned that he mixed his colors in pots rather than on a palette; see "Inner Monologue" (1959, 6).

70 Tom Learner, former conservation scientist at the Tate Gallery, London, analyzed the sample from *Pastorale* with pyrolysis-GC. He presented the results of analysis at a talk at the National Gallery of Art, Washington, D.C., on April 1, 1996.

71 When peeled off, the newspaper sheets often left reversed, offset images in the wet paint. Such an image is visible in the lower right corner of *Easter Monday* (1955–56, Metropolitan Museum of Art).

72 De Kooning also said that he "could sustain [the woman theme] all the time because it could change all the time"; see T. Hess (1965, 64). Harold Rosenberg called the new women "today's cuties" in his article for *Vogue* magazine (Rosenberg 1964, 187).

73 As early as 1960, de Kooning expressed his growing confidence: "I get freer, I feel I am getting more to myself in the sense of, I have all my forces... I am more convinced about picking up the paint and the brush and drumming it out"; see (Sylvester 1963, 48).

74 The paper supports of *Two Standing Women* and *Women* (1964) both have embossed Strathmore stamps.

75 The water resistance and absorbency of paper is controlled mainly with the addition of size to the paper. Calendering is a process for imparting a smooth, glossy finish to paper by passing it through a series of metal rollers. Vellum traditionally denotes animal skin, but it is now commonly used to refer to any treated tracing paper made from wood fibers or cotton.

76 Brockman reported that "not much was cooking for him at that point. No one quite knew what to make of it [his new nudes] or where he was at. Because he was in a major transition, as was the art scene."

77 The resin was identified using FTIR spectroscopy. Acrylic resins first became commercially available in the United States during the 1930s. Aerosol paints were introduced in the early 1950s, and aerosol varnish was commercially available by the early 1960s. Information supplied by Carlene Stephens, National Museum of American History, Smithsonian Institution. "Orange-peel" formation is the development of a dimpled pattern on the surface of paint, lacquer, or varnish. It is due to flow of an excessively thick coating applied to a vertical surface; see M. Hess (1965, 75).

78 *Two Standing Women* is the only work examined by the author that makes such an extensive use of a visible synthetic resin layer. This appears to have been an experimental procedure that the artist explored and then abandoned.

79 Paint cross-sections taken from the edges of the painting all show a resin layer between the support and the paint layers. As the resin was very thin and almost entirely covered by paint, it was not conclusively identified.

80 According to McMahon (1993), de Kooning tended to delineate the female body or parts of the body with the least possible number of strokes; see Shiff (1994, 57). Louis Finkelstein (1967, 202–206). compared de Kooning's abbreviated treatment of the figure and its frankly sexual content to graffiti.

81 De Kooning told his friend Rose Slivka that his two great loves of life were painting and his enjoyment of women (Slivka 1989, 220).

82 As early as 1959, Thomas Hess noted that "the theme of sexuality... is a major part of de Kooning's painting" (T. Hess 1959b, 30; 1972a, 229).

83 Art historians have frequently discussed de Kooning's women of the 1960s as an extension of the tradition of images of Titian, Rubens, and Renoir. More recently, Cooke and Shiff explored the deeper cultural, philosophical, and psychological roots of de Kooning's aesthetics. See Cooke (1993) and Shiff (1994).

84 Analysis of the paints in *Sphinx* reveals that the pigments as well as the binding medium of this work are the same as those of *Woman, Sag Harbor*, indicating that this work was probably among the many oil transfers that de Kooning made from one of the door paintings (see the next section of this chapter). A wood-grain pattern impressed into the paint surfaces of *Sphinx* as well as wood fibers caught in the paint surfaces provide further evidence that this work is a transfer from a wooden surface, possibly one of the doors, before a ground was applied.

85 Dealer Harold Diamond convinced de Kooning to show these works. A 1965 exhibition at the Paul Kantor Gallery in Los Angeles featured these transfers, called "monoprints."

86 There are several small works on transparent sheets of paper in the Hirshhorn Museum collection that have charcoal on one side and paint on the other.

87 De Kooning first experimented with lithography in 1960 at the University of California, Berkeley. In 1966 and 1970–71, in collaboration with Irwin Hollander, he explored this process again; see Graham (1991a and 1991b).

88 Collectors Kimiko and John Powers, who acquired the related painting *Singing Woman* (1965), confirmed McMahon's statement; see Kimiko and John Powers's letter to Judith Zilczer, May 4, 1992, in Zilczer (1993b, 161, n. 14). It was previously assumed that de Kooning purchased the doors specifically as supports for his paint-

ings; see T. Hess (1965, 37). Although John McMahon remembers that as many as six doors were painted, there are only three extant paintings on doors: *Woman, Sag Harbor*; *Woman* (1965); and *Woman Springs* (Davis Museum and Culture Center, Wellesley College). *Woman* (1964, Hirshhorn Museum and Sculpture Garden); *Woman Accabonac* (1966, Whitney Museum of American Art); and *Singing Woman* (private collection) are on paper mounted to canvas or to panel. Whether painted on wood or on vellum paper, these works also are called "door paintings" because they share the vertical format and have similar dimensions. The works on vellum probably resulted as transfers taken from an image of one of the door paintings, which were then reworked. For further discussion of this series, see Fox (1995).

89 Thomas Hess (1972c, 43) observed that de Kooning sometimes scraped his paints down using paint remover.

90 Regarding *Woman Accabonac*, de Kooning said, "She's kind of funny-looking; she's here for the weekend.... Well, she's turning into her forties, I guess. It's true, isn't it? She's sweet and friendly. Don't you think so?" (Wolfe 1981, 14).

91 See also Shiff (1994, 35). Although the sequence of paintings that were generated from paintings and oil transfers was seldom obvious, de Kooning enjoyed telling visitors to his studio which painting came from which; see Wolfe (1981, 15–16).

92 Richard Shiff (1994, 37) identified this jumping pose with a newspaper photograph of cheerleaders that de Kooning had hanging in his studio and with popular images of fashion models.

93 De Kooning also told Rosenberg (1972, 57) that to make these colors, which he called "tones," all he needed was "white and orange to give me an ideal, bizarre flat color." De Kooning named his colors after real things: tree color, sky color, skin color, according to Charlotte Willard (1969a, 56). There are more than thirty undated annotated swatches of the artist's different color mixtures in the collection of The Willem de Kooning Foundation, New York. With the exception of the browns and blacks, few of these color formulations are complex. Most of them specify that one, two, and sometimes three tubes of colors be combined with several tubes of white.

94 De Kooning said he made his colors by mixing what he called "mineral colors." He worked out this system because these colors remained wet longer than the earth pigments and they did not "sink in" or dry unevenly the way earth colors did (Shirey 1967, 80). Technically, mineral pigments are those derived from the natural minerals of the earth, including the earth pigments that de Kooning said he was avoiding. In using this term, he was expressing his empirical response to his materials rather than describing with technical accuracy their compositions and properties. Thus, he probably was referring to the relatively new synthetic pigments, such as the cadmiums and titanium white, which are so commonly identified in the paintings of this period.

95 The prevalence of synthetic organic pigments and dyes in de Kooning's paintings of the 1960s, many of them fluorescent under ultraviolet light, merits further research. Organic pigments are often mixed with large quantities of extenders or fillers, and unfortunately they are difficult to identify in small samples; see Berrie and Lomax (1997). A strongly fluorescent sample of pink paint from *Woman* (1965) was identified by Suzanne Lomax, organic chemist, as a rhodamine dye (Analysis Report, February 7, 1997, Scientific Research Department, National Gallery of Art). A sample of yellow paint from the top right corner of *Woman* (1965) was identified by the author as cadmium yellow in combination with one of the arylide yellows.

96 In "Inner Monologue" (1959, 6), de Kooning called his prepared paints "pots of light."

97 No house paints have been identified in any of the paintings executed after 1963. Although McMahon (1993) recalls that many cans of house paint were moved from the New York studio to Long Island, he does not remember ever seeing de Kooning use them.

98 Saff-o-life oil was a General Foods product. Safflower oil was introduced into the United States around 1955; see Martens (1974, 25–26).

99 Shaking the emulsified paint in a jar, de Kooning claimed "it only costs 79 cents a bottle"; see Dickerson (1964a, 6).

100 De Kooning recorded the ingredients of his mixtures on a handmade chart that can be seen hanging on the wall to the left of the easel in photographs of the artist's studio. The whites in the paintings of the early 1960s are generally a mixture of titanium dioxide and zinc oxide pigments with large amounts of barium sulfate. Bocour's oil paint formulations show that Bellini Titanium White was a mixture of 2.0 parts titanium dioxide, 0.5 parts zinc oxide, and 1.5 parts blanc fixe (The Artists' Materials Research and Study Center, National Gallery of Art, Washington, D.C.). At some later date, probably in the early 1980s, de Kooning switched from Bellini oils to Winsor and Newton paints.

101 De Kooning's description of Bocour's safflower oil as the "right kind of oil" is interesting. While safflower oil is regularly used as a binding medium in artists' paints, especially with white pigments, driers are added to accelerate the slower drying rate of this oil; Gamblin (1998). Undated recipes for Bocour's oil paint formulations in the Artists' Materials Research and Study Center at the National Gallery of Art, Washington, D.C., show that both Bellini Titanium White and Flake White were ground in safflower and castor oil with the addition of large amounts of driers. Thus, Bocour may have explained to de Kooning—who by the 1970s and possibly as early as the 1960s was having difficulty with the slow drying rate of his paints—that nondrying and semi-drying oils required treatment with driers to ensure that they would set in a reasonable time period, especially in the presence of inert pigments; see Chapter 4 for further discussion of this issue.

102 Naturally aged three-year-old standards of pigmented and unpigmented safflower oil had P/S ratios from 2.5 to 3.2, with a mean of 2.8. The paint mock-ups were prepared from artists' tube paints ground in linseed oil, to which safflower cooking oil was added in 10%, by weight, increments. The pigments present in the standards were a mixture of cadmium red, titanium dioxide, zinc oxide, and barium sulfate, roughly according to percentages found in those in de Kooning's paintings of the 1960s. The paints were applied to glass plates and were stored at ambient temperature and humidity for two years. The paints with between 50 and 100%, by weight, of added safflower oil had P/S ratios ranging from 2.2 to 3.3.

103 Many of the paints in the works executed by about 1964 have remained soft or gooey over forty years after they were completed. Soft paints, in fact, are a common characteristic of de Kooning's paintings well into the 1970s. See Chapter 4 for a discussion of the condition of de Kooning's paintings from the 1960s and 1970s.

104 The rumor that de Kooning added mayonnaise to his paints was widespread orally and was perpetuated in print in Waldman (1978, 26); Russell (1978, sect. 2, 25); and Anfam (1990, 180). In interviews, de Kooning vehemently denied such reports; see Rodgers (1978, 20) and Cherry (1989, 230).

105 During a 1964 interview, de Kooning also said, "I've been working on this new painting for the last two months, changing it two or three times every day. I've scooped it off a lot. There must be twenty paintings under it"; see Dickerson (1964b, 70).

106 The pigment mixture present in this prepared paint sample was similar to that used by de Kooning: cadmium orange combined with a white (1:20 by weight), which was a mixture of titanium dioxide, zinc oxide, and barium sulfate.

107 Peter Rudolph, telephone conversation, March 8, 2002.

108 De Kooning made a similar statement about the importance of puddles for his imagery in a *Time* magazine article (1959). Likewise, in a 1963 interview, he described how he wanted to make the whole painting "look like a [bottle], like a lot of [bottles], where everything is very light"; see Sylvester (1963, 48). The word "bottle" was probably transcribed incorrectly and should be "puddle"; see Prather (1994, 181 n. 39).

109 Lucy Lippard (1965, 31) noted that in the works of this period the woman's "stability has vanished [and the] frothy, loose pigment with its drips, blots . . . and bubbles stresses movement rather than structure."

110 In a 1957 interview with Irving Sandler (Sandler 1989, 216), de Kooning often referred to himself as a "conservative" who worked under the influence of European modernism.

References

Alloway, Lawrence. 1960. Sign and surface: Notes on black and white painting in New York. *Quadrum* 9, 49–62.

———. 1961. Iconography wreckers and maenad hunters. *Art International* 5 (April): 32–47.

———. 1965. The biomorpic forties. *Art Forum* 4 (September): 18–23.

———. 1975. De Kooning: Criticism and art history. *Art Forum* 13 (January): 46–50.

Anfam, David. 1990. *Abstract Expressionism*. London: Thames and Hudson.

Arb, Renée. 1948. Spotlight: De Kooning. *ARTnews* 47 (April), 33.

Artists' Materials Research and Study Center, National Gallery of Art, Washington, D.C.

Ashton, Dore. 1955. Art. *Art and Architecture* 72 (December), 34.

———. 1965. *Willem de Kooning*. Northampton, Mass.: Smith College Museum of Art.

———. 1968. New York commentary: De Kooning at Knoedler. *Studio International* 175 (January), 39–41.

———. 1969. New York commentary: de Kooning at the Museum of Modern Art and Knoedler. *Studio International* 177 (May), 243–45.

———. 1972. *The New York School: A Cultural Reckoning*. New York: Viking.

Auping, Michael, ed. 1987. *Abstract Expressionism: The Critical Developments*. Buffalo: Alright-Knox Art Gallery.

Barclay, Marion H. 1992. Materials used in certain Canadian abstract paintings of the 1950s. In *The Crisis of Abstraction in Canada: The 1950s*, by Denise Le Clerc, 205–32. Ottawa: National Gallery of Canada.

Barr, Alfred, Jr. 1950. Gorky, de Kooning, and Pollock. *ARTnews* 49 (Summer), 22, 60.

Berkson, Bill, and Rackstraw Downes, eds. 1989. Willem de Kooning, on his eighty-fifth birthday. *Art Journal* 48 (Fall), entire issue devoted to de Kooning.

Berman, Avis. 1982. Willem de Kooning: "I am only halfway through." *ARTnews* 81 (February), 68–73.

Berrie, Barbara, and Suzanne Quillen Lomax. 1997. Azo pigments: Their history, synthesis, properties, and use in artists' materials. *Conservation Research 1996/1997*. Washington, D.C.: National Gallery of Art.

Boon, Jaap J., Jaap van der Weerd, Katrien Keune, Petria Noble, and J. Wadum. 2002. Mechanical and chemical changes in Old Master paintings: Dissolution, metal soap formation and remineralization processes in lead pigmented ground/intermediate paint layers of 17th century paintings. *Preprints: ICOM Committee for Conservation, 13th Triennial Meeting, Rio de Janeiro, 22–27 September 2002*, ed. Roy Vontobel, 1: 401–6.

Bourdrez, Martha. 1950. De Kooning: Painter of promise: A Dutch rebel on his way to the top. *Knickerbocker* 12 (May), 8–9.

Bozo, Dominique, Claire Stoullig, et al. 1984. *Willem de Kooning*. Paris: Musée national d'art moderne, Centre George Pompidou.

Brach, Paul. 1951. Fifty-Seventh Street in review. *Art Digest* 26 (December), 19.

Brady, George S., and Henry R. Clauser. 1977. *Materials Handbook: An Encyclopedia for Managers, Technical Professionals, Purchasing and Production Managers, Technicians, Supervisors, and Foremen*. 11th ed. New York: McGraw-Hill.

Bürgi, Bernhard Mendes, Klaus Kertess, and Ralph Ubl. 2006. *De Kooning: Paintings, 1960-1980*. Basel:Kunstmuseum.

Butler, Cornelia, and Paul Schimmel, eds. 2002. *Willem de Kooning: Tracing the Figure*. Los Angeles: Museum of Contemporary Art.

Carmean, E.A., Jr. 1978. Willem de Kooning: The women. In *American Art at Mid-Century: The Subjects of the Artist*, by Carmean and Eliza Rathbone, 154–82. Washington, D.C.: National Gallery of Art.

Cateforis, David. 1992. Willem de Kooning's "women" of the 1950s: A critical history of their reception and interpretation. Ph.D. diss. Stanford University. Ann Arbor, Mich.: University Microfilms.

———. 1994. *Willem de Kooning*. New York: Rizzoli.

Cherry, Herman. 1989. Willem de Kooning. *Art Journal* 48 (Fall), 230.

Cooke, Lynne. 1993. De Kooning and the pastoral: The interrupted idyll. In *Willem de Kooning from the Hirshhorn Museum Collection*, 89–109. Washington, D.C.: Hirshhorn Museum and Sculpture Garden (HMSG), Smithsonian Institution.

Cowart, Jack. 1979. De Kooning today. *Art International* 23 (Summer), 8–17.

Cowart, Jack, and Sanford Sivitz Shaman. 1978. *De Kooning 1969–1978*. Cedar Falls: University of Northern Iowa Gallery of Art.

Cummings, Paul. 1978. Interview with Leonard Bocour. Transcript of a taped interview, New York, June 8. Archives of American Art, Smithsonian Institution, Washington, D.C.

———. 1983. The Drawings of Willem de Kooning. In *Willem de Kooning: Drawings, Paintings, Sculpture*. New York: Whitney Museum of American Art.

De Antonio, Emile, and Mitch Tuchman. 1984. *Painters Painting: A Candid History of the Modern Art Scene, 1940–1970*. New York: Abbeville.

De Kooning, Elaine. 1983. De Kooning memories. *Vogue* (December), 352–53, 393–94.

De Kooning, Willem. 1949a. Editors' letters [on Arshile Gorky]. *ARTnews* 47 (January), 6.

———. 1949b. A desperate view. Talk delivered at The Subjects of the Artist: A New School, New York, February 18, 1949. Transcribed in T. Hess 1968, 15–16.

———. 1951a. The Renaissance and order. Talk delivered at Studio 35. In *Trans/formation* 1, 85–87.

———. 1951b. What abstract art means to me. Paper prepared for a symposium at the Museum of Modern Art, New York. In *Museum of Modern Art Bulletin* 18 (Spring), 4–8.

———. 1988. *The Collected Writings of Willem de Kooning*, ed. George Scrivani. Madras and New York: Hanuman.

Denby, Edwin. 1964. My friend de Kooning. *ARTnews Annual* 29, 82–99, 156.

———. 1988. *Willem de Kooning*. New York: Hanuman. Includes reprints of poems "Willem de Kooning" and "The Thirties."

De Wilde, Edy, and Carter Ratcliff. 1983. *Willem de Kooning: Het Noordatlantisch Licht/The North Atlantic Light, 1960–1983*. Amsterdam: Stedelijk Museum.

Dickerson, George. 1964a. Transcript of an interview with Willem de Kooning, September 3, 1964. On file with the Thomas B. Hess Papers, Archives of American Art, Smithsonian Institution, Washington, D.C.

———. 1964b. The strange eye and art of de Kooning. *Saturday Evening Post* (November 21), 68–71.

Doerner, Max. 1984. *The Materials of the Artist*. New York: Harcourt.

Feller, Robert L., ed. 1986. *Artists' Pigments: A Handbook of Their History and Characteristics*. 3 vols. Washington, D.C.: National Gallery of Art.

Finkelstein, Louis. 1950. Marin and de Kooning. *Magazine of Art* 43 (October), 202–6.

———. 1967. The light of de Kooning. *ARTnews* 66 (November), 28–31, 70–71.

Fitzsimmons, James. 1953. Art. *Arts and Architecture* 70 (May), 4, 6–10.

Forge, Andrew. 1968. De Kooning's "Women." *Studio International* 176 (December), 246–251.

———. 1969. De Kooning in retrospect. *ARTnews* 68 (March), 44–47, 61–63.

Fox, Judith Hoos. 1995. *Willem de Kooning's Door Cycle*. Wellesley, Mass.: Davis Museum and Cultural Center, Wellesley College.

Frasnay, Daniel. 1969. *The Artist's World*. London and New York: Viking.

Friedman, Ted (technical consultant for Insel-X Products, Yonkers, N.Y.). 1992. Telephone communication with Susan Lake, September 3. Willem de Kooning Conservation Files, HMSG, Smithsonian Institution.

Gamblin, Robert (owner, Gamblin Artists Colors Co.). 1998. Telephone conversation with Susan Lake, September 10. Willem de Kooning Conservation Files, HMSG, Smithsonian Institution.

Garrels, Gary. 1995. Three toads in the garden: Line, color, and form. In *Willem de Kooning: The Late Paintings, the 1980s*, ed. Gary Garrels and Robert Storr, 9–37. San Francisco: San Francisco Museum of Art.

Gaugh, Harry F. 1983. *Willem de Kooning*. New York: Abbeville.

Geist, Sidney. 1953. Work in Progress. *Art Digest* 27 (April 1), 15.

Geldzahler, Henry. 1987. *Willem de Kooning: Abstract Landscapes, 1955–1963*. New York: Gagosian Gallery.

Goldwater, Robert. 1960. Reflections on the New York School. *Quadrum* 8 (1960), 17–36.

Gornitzky, Jay (President, Empire Blended Products, Bayville, N.J.). 1998. Telephone conversation with Susan Lake, September 10, Willem de Kooning Conservation Files, HMSG, Smithsonian Institution.

Graham, Lanier. 1991a. *The Prints of Willem de Kooning: A Catalogue Raisonné 1957–1971*. Vol. 1. Paris: Baudoin Lebon.

———. 1991b. *Willem de Kooning: Printer's Proofs from the Collection of Irwin Hollander, Master Printer*. New York: Salander-O'Reilly Galleries.

Greenberg, Clement. 1940. Toward a newer Laocoon. *Partisan Review* (July–August), 296–310.

———. 1948a. Art. *Nation* 166 (January 24), 107–08.

———. 1948b. Art. *Nation* 166 (April 24), 448.

———. 1955. "American-type" painting. *Partisan Review* 22 (Spring), 179–96.

Heaton, Noël. 1947. *Outlines of Paint Technology*. 3rd ed. London: Charles Griffin.

Hess, Barbara. 2004. *De Kooning (1904–1997): Content as a Glimpse*. Köln: Taschen.

Hess, Manfred, ed. 1965. *Paint Film Defects, Their Causes and Cure*. 2nd ed. New York: Reinhold.

Hess, Thomas B. 1953. De Kooning paints a picture. *ARTnews* 52 (March), 30–33, 64–67.

———. 1958. Is today's artist with or against the past? *ARTnews* 57 (June), 27, 56.

———. 1959a. Reviews and Previews. *ARTnews* 58 (May), 12–13.

———. 1959b. *Willem de Kooning*. New York: George Braziller.

———. 1962. *Recent Paintings by Willem de Kooning*. New York: Sidney Janis Gallery.

———. 1965. De Kooning's new women. *ARTnews* 64 (March), 36–38, 63–65.

———. 1967. *De Kooning: Recent Paintings*. New York: Walker and Company for M. Knoedler.

———. 1968. *Willem de Kooning*. New York, Museum of Modern Art.

———. 1972a. Pinup and icon. In *Woman as Sex Object 1730–1970*, ed. Thomas B. Hess and Linda Nochlin. *ARTNews Annual* 38 (1972), 222–37.

———. 1972b. Art: The flying Hollander of Easthampton, L.I. *New York Magazine* (February 20), 68–72.

———. 1972c. *Willem de Kooning: Drawings*. Greenwich, Conn.: NY Graphic Society.

———. 1975. Art: Water babes. *New York Magazine* (October 27), 73–75.

———. 1978. In de Kooning's studio. *Vogue* (April), 234–37, 335.

Hopkins, Budd. 1984. The drawings of Willem de Kooning. *Drawing* 5 (March–April), 121–25.

Hunter, Sam. 1975. De Kooning: 'Je dessine les yeux fermés.' *Galerie Jardin des Arts* 152 (November), 68–70.

"Inner Monologue." 1959. Transcript of a taped discussion with Willem de Kooning, Michael Sonnabend, and Kenneth Snelson, Summer 1959. Willem de Kooning Office, New York. Parts of the dialogue are incorporated into Robert Snyder's two films, *Sketchbook No. 1* (1960) and *A Glimpse of de Kooning* (1968).

International Correspondence Schools. 1899. *Elements of Lettering and Sign Painting*. Scranton, Pa.: International Textbook Company.

Janis, Harriet, and Rudi Blesh. 1960. *De Kooning*. New York and London: Grove.

Kertess, Klaus. 2000. *Willem de Kooning in Process*. Fort Lauderdale: Museum of Art.

Kleeblatt, Norman L. 2008. *Action/Abstraction: Pollock, de Kooning, and American Art, 1940–1965*.New York: Jewish Museum.

Koller, Johann, and Andreas Burmester. 1990. Blanching of unvarnished modern paintings: A case study on a painting by Serge Poliakoff. In *Cleaning, Retouching and Coatings: Technology and Practice for Easel Paintings and Polychrome Sculpture: Preprints of the Contributions to the Brussels Congress, 3–7 September 1990*, ed. John S. Mills and Perry Smith, 138–143. London: International Institute for Conservation of Historic and Artistic Works.

Kramer, Hilton. 1973. Willem de Kooning. In *The Age of the Avant-Garde: An Art Chronicle of 1956–1972*, 347–53. New York: Farrar,

Straus and Giroux.

———. 1994. The ghost of de Kooning. *Modern Painters* 7 (Summer), 32–35.

Lader, Melvin P. 1978. Graham, Gorky, de Kooning, and the "Ingre Revival" in America. *Arts Magazine* 52 (March), 94–99.

Lake, Susan, and Judith Zilczer. 1993. Painter and draftsman. In *Willem de Kooning from the Hirshhorn Museum Collection*, 172–79. Washington, D.C.: HMSG, Smithsonian Institution.

Larson, Philip, and Peter Schjeldahl. 1974. *De Kooning: Drawings/Sculptures*. Minneapolis: Walker Art Center.

Lerner, Abram. 1975. A discussion with Willem de Kooning, Springs, East Hampton, L.I., November 24, 1975. Typescript. Collection Archive, HMSG, Smithsonian Institution.

———. 1977. *De Kooning*. Belgrade: USIA and Museum of Contemporary Art.

Life. 1951. The Metropolitan and modern art (January 15), 34–38.

Lippard, Lucy R. 1965. Three generations of women: De Kooning's first retrospective. *Art International* 9 (November), 29–31.

Louchheim, Aline B. 1951. Six Abstractionists defend their art. *New York Times Magazine* (January 21), 16–17.

Martens, Charles R., ed. 1974. *Technology of Paints, Varnishes and Lacquers*. Huntington, N.Y.: R.E. Krieger.

Mason, Michael E., Mary E. Eager, and George R. Waller. 1964. A procedure for the simultaneous quantitative determination of glycerol and fatty acid contents of fats and oils. *Analytical Chemistry* 36 (3), 597–90.

Mayer, Ralph. 1970. *The Artist's Handbook of Materials and Techniques*. New York: Viking.

———. 1981. *The Artist's Handbook of Materials and Techniques*. New York: Viking.

McBride, Henry. 1953. Abstract report for April. *ARTnews* 52 (April), 47.

McMahon, John (artist). 1993. Personal communication with Susan Lake, October 18–21. Willem de Kooning Conservation Files, HMSG, Smithsonian Institution.

Michalski, Stefan. 1990. A physical model for the cleaning of oil paint. In *Cleaning, Retouching and Coatings: Technology and Practice for Easel Paintings and Polychrome Sculpture: Preprints of the Contributions to the Brussels Congress, 3–7 September 1990*, ed. John S. Mills and Perry Smith, 85–92. London: International Institute for Conservation of Historic and Artistic Works.

Mills, John. 1966. The gas-chromatographic examination of paint media. Part I. Fatty acid composition and identification of dried oil films. *Studies in Conservation* 11, 92–107.

Mills, John, and Raymond White. 1977. Natural resins of art and archaeology: Their sources, chemistry, and identification. *Studies in Conservation* 22, 12–31.

———. 1980. Analyses of paint media. *National Gallery Technical Bulletin* 4, 65–7.

———. 1994. *The Organic Chemistry of Museum Objects*. 2nd ed. Oxford and Boston: Butterworth-Heinemann.

Moffat, Elizabeth A., and David Miller. 1996. Spectroscopic and chromatographic analysis of selected paintings from the Parisian period of Paul-Emile Borduas. *Journal of the International Institute of Conservation - Canadian Group* 19, 3–13.

Mooradian, Karlen. 1971. Interview with Willem de Kooning. Springs, East Hampton, July 19, 1966. In Remembrances of Gorky. *Ararat* 12 (Fall), 48–52.

Newsweek. 1968. The native returns (October 7), 116.

———. 1969. De Kooning's year (March 10), 91.

O'Connor, Francis, and Eugene Victor Thaw. 1978. *Jackson Pollock: A Catalogue Raisonné of Paintings, Drawings, and Other Works*. Vol. 4. New Haven and London: Yale University Press.

Parsons, Anne Bowden. 1967–68. Interview with de Kooning. Transcript of an interview, October 11, 1967. Anne Bowden Parson's Collection of Interviews on Art. Archives of American Art, Smithsonian Institution, Washington, D.C.

Passlof, Patricia (artist). 1998. Telephone conversation with Susan Lake, August 4, 1998. Willem de Kooning Conservation Files, HMSG, Smithsonian Institution.

Patton, Temple C., ed. 1973. *Pigment Handbook*. Vol. 1, *Properties and Economics*. New York: Wiley.

Pepper, Curtis Bill. 1983. The indomitable de Kooning. *New York Times Magazine* (November 20), 42–47, 66, 70, 86, 90, 94.

Perreault, John. 1969. De Kooning at the Springs. *New York Magazine* (March 17), 44–47.

Polcari, Stephen. 1991. *Abstract Expressionism and the Modern Experience*. Cambridge: Cambridge University Press.

Powder Diffraction File. 1997. International Centre for Diffraction Data. http://www.icdd.com

Prather, Marla. 1994. Catalogue. In *Willem de Kooning: Paintings*, 75–224. Washington, D.C.: National Gallery of Art.

Rodgers, Gaby. 1978. Willem de Kooning: The artist at 74. *LI: Newsday's Magazine for Long Island* (May 21), 18–20, 35–36.

Rodman, Selden. 1961. *Conversations with Artists*. New York: Capricorn.

Rose, Barbara. 1978. Hans Hofmann: From Expressionism to abstraction. *Arts Magazine* 53 (November), 110–14.

Rosenberg, Harold. 1952. The American action painters. *ARTnews* 51 (December), 22–23, 48–50.

———. 1962. Action painting: A decade of distortion. *ARTnews* 61 (December), 42–44, 62–63.

———. 1963. The art galleries: Painting is a way of living. *New Yorker* (February 16), 126, 128, 130–37. Reprinted in Harold Rosenberg (1966), *The Anxious Object: Art Today and Its Audience*, 109–20. New York: Horizon.

———. 1964. De Kooning. *Vogue* (September), 146–49, 186–87.

———. 1972. Interview with Willem de Kooning *ARTnews* 71 (September), 54–59.

———. 1974. *Willem de Kooning*. New York: Abrams.

———. 1978. The art world: Woman and water. *New Yorker* (April 24), 132–35.

Rosenblum, Robert. 1990. Notes on de Kooning. In *Willem de Kooning: An Exhibition of Paintings*. New York, Salander-O'Reilly Galleries.

Rosenzweig, Phyllis. 1980. *The Fifties: Aspects of Painting in New York*. Washington, D.C.: HMSG, Smithsonian Institution.

Rudolph, Peter (art collector and dealer). 2002. Personal communication with Susan Lake, March 8, Willem de Kooning Conservation Files, HMSG, Smithsonian Institution.

Russell, John. 1978. De Kooning: "I see the canvas and I begin." *New York Times* (February 5), sec. 2, 1, 25.

Sandler, Irving. 1965. The Club. *Art Forum* 4 (September), 27–31.

———. 1978. *The New York School: The Painters and Sculptors of the Fifties*. New York: Harper & Row.

———. 1989. Conversations with de Kooning, *Art Journal* 48 (Fall), 216–17.

———. 1993. Abstract Expressionism: The noise of traffic on the way to Walden Pond. In *American Art in the 20th Century: Painting and Sculpture 1913–1993*, 77–83. Berlin: Zeitgeist-Gesellschaft; London: Royal Academy of Arts.

Schilling, Michael R. 2005. Paint Media Analysis. In *Arthur M. Sackler Colloquium on the Scientific Examination of Art*, 186–205. Washington, D.C.: National Academy of Sciences.

Schilling, Michael R., and Herant P. Khanjian. 1996. Gas chromatographic determination of the fatty acid and glycerol content of lipids, Part I. The effects of pigments and aging on the composition of oil paints. In *Preprints: ICOM Committee for Conservation, 11th Triennial Meeting, Edinburgh, Scotland, 1–6 September, 1996*, ed. Janet Bridgland, 1:220–27. London: James & James.

Schilling, Michael R., David M. Carson, and Herant P. Khanjian. 1997. Gas chromatographic determination of the fatty acid and glycerol content of lipids, Part II. Fatty acid and glycerol content of lipids; effects of ageing and solvent extraction on the composition of oil paints. *Techne* 5 (1997), 71–78.

Schlenk, Hermann, and Joanne L. Gellerman. 1960. Esterification of fatty acids with diazomethane on a small scale. *Analytical Chemistry* 32, 1412.

Seitz, William C. 1962. *Arshile Gorky: Paintings, Drawings, Studies*. New York: Museum of Modern Art.

Shiff, Richard. 1987. Performing an appearance: On the surface of Abstract Expressionism. In *Abstract Expressionism: The Critical Developments*, 94–123. Buffalo: Albright-Knox Art Gallery.

———. 1994. Water and lipstick: De Kooning in transition. In *Willem de Kooning: Paintings*, 33–74. Washington, D.C.: National Gallery of Art.

Shirey, David L. 1967. Don Quixote in Springs. *Newsweek* (November 20), 80.

Slivka, Rose C. S. 1989. Willem de Kooning. *Art Journal* 48 (Fall), 219–21.

"Soirée." 1959. Transcript of a taped discussion with Willem de Kooning, Michael Sonnabend, Robert Snyder, Harold Rosenberg, Herman Cherry, Thomas B. Hess, and Franz Kline (July). Willem de Kooning Office, New York. Parts of dialogue are incorporated into Robert Snyder's two films, *Sketchbook No.1* (1960) and *A Glimpse of de Kooning* (1968).

Staats, Margaret, and Lucas Matthiessen. 1977. The genetics of art: Willem de Kooning. *Quest* 77 (March–April), 70–71.

Stevens, Mark. 1994. The master of imperfection. *New Republic* (July 4), 27–33.

———, and Annalyn Swan. 2004. *De Kooning: An American Master*. New York: Knopf.

Stone, Allan (art collector and dealer). 1998. Telephone conversation with Susan Lake, March 21, 1998. Willem de Kooning Conservation Files, HMSG, Smithsonian Institution.

Stuckey, Charles F. 1980. Bill de Kooning and Joe Christmas. *Art in America* 68 (March), 67–79.

Sylvester, David. 1963. "Content is a glimpse..." *Location* 1 (Spring), 45–53. Partial publication without interviewer's questions of "Painting as self-discovery," BBC broadcast of an interview with de Kooning aired December 30, 1960. Excerpts published with questions in Sylvester 1968 and 1969.

———. 1968. De Kooning's women, *Sunday Times Magazine* (London) (December 8), 44–52, 57.

———. 1969. De Kooning's women, *Ramparts* 7 (April), 20–24.

———. 1994. Flesh was the reason. In *Willem de Kooning: Paintings*, 15-31. Washington, D.C., National Gallery of Art.

Thomas Register of American Manufacturers 1950–70. New York: Thomas Publishing Company.

Thompson, David. 1969. The end of a tradition? *New York Times* (January 19), sec. 2, 25.

Time. 1951. Willem the walloper (April 30), 63.

Time. 1956. The wild ones (February 20), 70–75.

Time. 1959. Big splash (May 18), 72.

Tomkins, Calvin. 1987. Profiles: Colored muds in a sticky substance. *New Yorker* (March 16), 44–46, 48–55, 58–70.

Tworkov, Jack. 1950. The wandering Soutine. *ARTnews* 49 (November), 30–33, 62.

Ullman's Encyclopedia of Industrial Chemistry. 1985. 5th ed. Deerfield Beach, Fla.: VCH.

Valliere, James T. 1967. De Kooning on Pollock: An interview. *Partisan Review* 34 (Fall), 603–5.

Van den Berg, Jorrit Dirk Jan. 2002. *Analytical Chemical Studies on Traditional Linseed Oil Paints*. MOLART Report (University of Amsterdam). London: Archetype.

Waldman, Diane. 1978. *Willem de Kooning in East Hampton*. New York: Solomon R. Guggenheim Museum.

———. 1988. *Willem de Kooning*. New York: Abrams.

White, Raymond. 1978. The application of gas-chromatography to the identification of waxes. *Studies in Conservation* 23, 57–68.

White, Raymond, and Jennifer Pilc. 1996. Analyses of paint media. *National Gallery Technical Bulletin* 17, 91–95, 101–103.

Wilkie, Ken. 2005. Portrait of a modern master. In *A Visit to Willem de Kooning*, by Eddy Posthuma de Boer and Ken Wilkie, 20–36. Rotterdam: Willem de Kooning Academy, Rotterdam University.

Willard, Charlotte. 1964. In the art galleries. *New York Post Magazine* (August 23), 44.

———. 1969a. De Kooning: The Dutch-born U.S. master gathers in the international honors. *Look* (May 27), 54–59.

———. 1969b. Willem de Kooning transcript. Research notes of an interview with Willem de Kooning conducted for *Look* magazine. Archives of American Art, Smithsonian Institution, Washington, D.C.

Williams, Morris (Technical Director, T.J. Ronan Paint Corp., Bronx, N.Y.). 1998. Telephone conversation with Susan Lake, September 10. Willem de Kooning Conservation Files, HMSG, Smithsonian Institution.

Williams, Scott R. 1989. Blooms, blushes, transferred images and mouldy surfaces: What are these distracting accretions on art works? In *Proceedings of the 14th Annual IIC-CG Conference, IIC-Canadian Group, Ottawa*, 65–84.

Wolfe, Judith. 1981. *Willem de Kooning: Works from 1951–1981*. East Hampton, N.Y.: Guild Hall Museum.

———. 1996. *The Young de Kooning: Early Life, Training and Work, 1904–1926*. Ph.D. diss. City University of New York. Ann Arbor, Mich.: University Microfilms.

Yard, Sally. 1978. Willem de Kooning's women. *Arts Magazine* 53 (November), 96–101.

———. 1986. *Willem de Kooning: The First Twenty-Six Years in New York, 1927–1952*. New York: Garland Publishing.

———. 1991. The angel and the *demoiselle*—Willem de Kooning's *Black Friday*. *Record of the Art Museum, Princeton University* 50, 2–25.

———. 1997. *Willem de Kooning*. New York: Rizzoli.

———. 2007. *Willem de Kooning: Works, Writings, and Interviews*. Barcelona: Ediciones Poligrafa.

Zilczer, Judith. 1993a. *Willem de Kooning from the Hirshhorn Museum Collection*. Washington, D.C.: HMSG, Smithsonian Institution.

———. 1993b. Artist and patron: An exchange of letters. In ibid., 151–71.

———. 1993c. De Kooning and Urban Expressionism. In ibid., 11–57.

———. 1995. *Woman, Sag Harbor* and the origin of the "Door Paintings." In *Willem de Kooning's Door Cycle*. Wellesley, Mass.: Davis Museum and Cultural Center, Wellesley College.

———. 1996. Artist and patron: The formation of the Hirshhorn Museum's Willem de Kooning Collection. *Journal of the History of Collections* 8, 117–25.

Index

Note: page numbers in italics refer to figures; page numbers followed by "t" refer to tables, while those followed by "n" refer to notes, with note number.

Acknowledgments

Susan F. Lake

Director of Collection Management/
Chief Conservator
Hirshhorn Museum and Sculpture Garden
Smithsonian Institution

Most of the analytical work presented in this book was carried out at the University of Delaware and at the Getty Conservation Institute (GCI). At the GCI, I wish to thank in particular, Michael Schilling, senior scientist, who assisted with the analysis and generously advised me throughout the duration of the project. From the GCI, gratitude is also due to Tim Whalen, director; Tom Learner, senior scientist; Alberto de Tagle, former group director; James Druzik, senior scientist; and Narayan Khandekar, former associate scientist, all of whom greatly encouraged and promoted this study. Likewise, I am extremely grateful to Gordon Nicol, former mass-spectrometrist at the Department of Chemistry and Biochemistry at the University of Delaware. His thoughtful suggestions and assistance with analysis were essential to the realization of this project. I also wish to thank Judith Zilczer, former curator at the Hirshhorn Museum and Sculpture Garden for her advice and support throughout this study. Many thanks are also due to Willem de Kooning's studio assistant, artist John McMahon, who spent several days examining and discussing the artist's paintings with me. Walter Hopwood, former chemist, Smithsonian Museum Conservation Institute (MCI), generously gave me access to the infrared spectrometer. Nagase Takako of the Geological Sciences Department of the University of Delaware, Melanie Feather and Ronald Cunningham of MCI, and Eugena Ordonez, former conservator at the Museum of Modern Art, New York, generously assisted in the SEM-EDS analysis. Chris Maines and Suzanne Lomax, conservation scientists at the National Gallery of Art, Washington, D.C., assisted with medium analysis. I am also deeply grateful to Jay Krueger, painting conservator at the National Gallery of Art, who sampled many of the paintings not in the Hirshhorn Museum collection. Jay also contacted the curatorial staffs and private owners who permitted us to take paint samples from their pictures. Without their generosity, this project would not have been possible. Many thanks are also due to The Willem de Kooning Foundation for providing valuable research and editorial assistance. Finally, this project owes a great deal to Cynthia Godlewski at the Getty Conservation Institute; the Getty Publications team, editors Tobi Kaplan and Beatrice Hohenegger, production coordinator Suzanne Watson, and designer Jeffrey Cohen, as well as copy editor Robin Ray, who all helped turn a dissertation into a beautiful and readable book.